The Ingenious
Victorians

For Lucie and Katie, my two amazing daughters

The Ingenious Victorians

Weird and Wonderful Ideas from the Age of Innovation

John Wade

First published in Great Britain in 2016 by
PEN AND SWORD HISTORY
an imprint of
Pen and Sword Books Ltd
47 Church Street
Barnsley
South Yorkshire S70 2AS

Copyright © John Wade, 2016

ISBN 978 1 47384 901 3

The right of John Wade to be identified as the author of this work has been asserted by him in accordance with the Copyright, Designs and Patents Act 1988.

A CIP record for this book is available from the British Library
All rights reserved. No part of this book may be reproduced or transmitted in any form or by any means, electronic or mechanical including photocopying, recording or by any information storage and retrieval system, without permission from the Publisher in writing.

Printed and bound in England
by CPI Group (UK) Ltd, Croydon, CR0 4YY

Typeset in Times New Roman by
CHIC GRAPHICS

Pen & Sword Books Ltd incorporates the imprints of Pen & Sword Archaeology, Atlas, Aviation, Battleground, Discovery, Family History, History, Maritime, Military, Naval, Politics, Railways, Select, Social History, Transport, True Crime, Claymore Press, Frontline Books, Leo Cooper, Praetorian Press, Remember When, Seaforth Publishing and Wharncliffe.

For a complete list of Pen and Sword titles please contact
Pen and Sword Books Limited
47 Church Street, Barnsley, South Yorkshire, S70 2AS, England
E-mail: enquiries@pen-and-sword.co.uk
Website: www.pen-and-sword.co.uk

Contents

Introduction to an Era 7

Chapter 1   The World's Biggest Glass Building 15
Chapter 2   The Great Exhibition 23
Chapter 3   The Crystal Palace Reborn 37
Chapter 4   The Great Victorian Way 46
Chapter 5   When *Daddy Long-Legs* Ruled the Waves 51
Chapter 6   James Wyld's Monster Globe 60
Chapter 7   The Biggest Camera in the World................................67
Chapter 8   The First Channel Tunnel 74
Chapter 9   England's Answer to the Eiffel Tower 83
Chapter 10   Under, Over and Across the Thames 96
Chapter 11   Shocking Cures for Every Ailment...............................118
Chapter 12   Photographic Detectives 127
Chapter 13   Frédéric Kastner's Exploding Organ 135
Chapter 14   Trains in Drains 141
Chapter 15   How to Avoid Being Buried Alive 151
Chapter 16   London's Great Stink 161
Chapter 17   How London Got the Needle 172
Chapter 18   Phonographs, Graphophones and Gramophones 180
Chapter 19   Who Really Invented the Cinema?...............................189
Chapter 20   The Age of the Velocipede 200

Chapter 21 Pneumatic Railways 214

Chapter 22 The Rise and Fall of Electric Submarines 222

Chapter 23 Building Big Ben 228

Chapter 24 The Amazing Clocks of Charles Shepherd....................236

Chapter 25 From Today, Painting is Dead...244

Chapter 26 Victorian Flying Machines 257

Chapter 27 The People Who Invented Christmas 266

Chapter 28 An Amazing Memorial and the End of an Era 274

Bibliography 281

Picture credits 282

Index 286

Introduction to an Era

Queen Victoria acceded to the throne on 20 June 1837, at the age of eighteen, and there she remained for a record-breaking reign that lasted a little under sixty-four years (a record only recently beaten by Queen Elizabeth II). Victoria died on 22 January 1901. The age to which the Queen gave her name was a time when the British Empire was at the height of its powers. She ruled more than 450 million people in lands that stretched across the globe to include Africa, India, Canada, the Caribbean, Australia and New Zealand, covering about one quarter of the world's land mass.

No wonder the Victorians achieved a level of self-confidence that led them to produce so many successful inventors, designers and radical thinkers – and not just in Britain. When discussing the Victorian age, it is often accepted as applying only to Britain or, at least, to the lands over which the Queen reigned. But the era in general was equally a hotbed of invention around the world, especially in America and Europe.

Successful inventions
It was an age of ingenuity, inventiveness and innovation that saw the birth of many of the concepts and inventions that are still with us today. Consider a few of those that began life in the Victorian era.

Telephones: Scottish-born Alexander Graham Bell was a professor at Boston University when he worked in his spare time to find a way to send speech over wires. He filed his patent in 1876, beating rival Elisha Gray to the Patent Office by just a few hours and so winning the right to make telephones for public sale.

Radio: Guglielmo Marconi, on hearing scientific evidence that invisible waves travelled through the air, set about building a transmitter and a receiver to harness those waves for sending wireless messages. Unable to obtain backing in his native Italy, he travelled to Britain and set up the world's first permanent wireless station at The Needles, on the Isle of Wight, in 1897. In early experiments, he exchanged wireless messages with a tugboat in nearby Alum Bay, and then with Bournemouth, 14 miles away. In 1898, the world's first paid for radio telegram was sent from here, and in 1899, information for *The Transatlantic Times*, the first newspaper produced at sea, was transmitted from this station by wireless telegraphy. Today, a monument to Marconi's work stands on the site of the old wireless station.

The monument to Marconi's work at The Needles, on the Isle of Wight.

INTRODUCTION TO AN ERA

Postage stamps: Until Victorian times, postage on letters was paid by the recipient. English school teacher Rowland Hill saw the inconvenience of this, especially if the recipient couldn't afford, or didn't want, to accept the letter, and so set about producing adhesive stamps, paid for by the sender. The first, launched in 1840, was black, cost one penny to send a letter anywhere in the country and bore an image of Queen Victoria. Pillar boxes erected for the posting of stamped letters were originally green before red, with the monarch's initials on the front, and became the standard in the 1870s.

The Penny Black postage stamp bore Queen Victoria's head and enabled letters to be posted anywhere in the country for one penny.

Computers: English mathematician Charles Babbage designed mechanical calculators, the pinnacle of which was his analytical engine, conceived in 1834. It was capable of addition, subtraction, multiplication and division, and could also be programmed using punched cards, inspired by similar devices used in the textile industry for 'programming' looms to weave complex patterns. It could store numbers and separate them from the mathematical processes, and output its results as hard copy, punched cards and graphs. Babbage's next innovation, which he called his difference engine, and which he designed between 1847 and 1849, could calculate numbers with up to thirty-one digits. Much of his pioneering work was later reflected in the basic theory behind the workings of today's computers.

Fibre optics: Fibre-optic technology, in which sounds are transmitted by light, became popular in the 1980s. But a century before, in 1880, telephone pioneer Alexander Graham Bell succeeded in transmitting speech from the roof of a nearby school to his laboratory, about 700 feet away, on a modulated beam of light. It worked by directing speech at a flexible mirror, which became convex or concave according to the sound. He called his invention the photophone.

Railways: It was the age of the steam railway, when many already existing small local railways were unified and integrated across the country. As railway fever gripped the nation, a standard gauge (the distance between the rails) was set, magnificent stations were built and amazing feats of engineering resulted in impressive bridges, cavernous tunnels and

THE INGENIOUS VICTORIANS

Still in daily use, the brick-built, 100ft high Welwyn Viaduct in Hertfordshire was opened by Queen Victoria in 1850.

vertiginous viaducts, a great many of which still survive and are in daily use.

Successful concepts
The Victorian age of ingenuity was not confined to inventions. Equally ingenious were the great ideas, plans and concepts that became reality.

Who but the Victorians would have thought of staging an enormous exhibition of world culture and industry, then building the world's largest glass structure to house it, before tearing the building down, transporting it piece by piece across London and rebuilding it at the centre of probably the world's first theme park?

Who else would dream of building an electric railway whose lines ran under the sea with the carriage supported on stilts 20 feet above the waves?

What other age could have given us long-lasting electric light bulbs, sound recording, an early attempt at a Channel tunnel, the first flying machines, steam trains under London, life-preserving coffins, four British copies of the French Eiffel Tower and the recipe for Coca-Cola? The Victorians also invented Christmas as we know it today.

Mistaken opinions
Not everyone who heard about the latest advances in ideas or technology had the foresight to understand that maybe they were on the brink of a discovery that might change the world. Here are some of the mistaken opinions put forward by eminent people of the day.

'Rail travel at high speeds is not possible because passengers, unable to breathe, would die of asphyxia.'
Dionysius Lardner, Professor of Natural Philosophy and Astronomy at University College, London, 1830.

INTRODUCTION TO AN ERA

'Your cigar-ettes will never become popular.'
Cigar-maker, F.G. Alton, turning down John Player, 1870.

'This "telephone" has too many shortcomings to be seriously considered as a means of communication. The device is inherently of no value to us.'
Western Union internal memo, 1876.

'The Americans have need of the telephone, but we do not. We have plenty of messenger boys.'
Sir William Preece, Chief Engineer, British Post Office, 1876.

'When the Paris Exhibition closes, electric light will close with it and no more will be heard of it.'
Oxford professor, Erasmus Wilson, 1878.

'The phonograph is not of any commercial value.'
Inventor, Thomas Edison, 1880.

'Radio has no future. Heavier-than-air flying machines are impossible. X-rays will prove to be a hoax.'
British scientist, Lord Kelvin, 1899.

'Everything that can be invented has been invented.'
Charles H. Duell, Commissioner, US Patent Office, 1899.

Heroic failures
Of course, not every invention or concept was a success. The Victorian age also had its fair share of failures.

In 1868, in an attempt to control the rising tide of London traffic, the world's first traffic lights were erected in Westminster. They took the form of twin semaphore-like arms – straight out meant stop, both pointing down meant caution – with red and green gas lamps above, operated manually by a policeman. A month after the lights were erected, they exploded, seriously injuring the policeman, and deemed too dangerous to remain, they were removed. Today, in honour of inventor John Peake Knight's revolutionary, if somewhat dangerous, precursor to today's traffic lights, a commemorative plaque can be seen on the side of a building in Bridge Street, Westminster, opposite the Houses of Parliament.

THE INGENIOUS VICTORIANS

In 1883, a patent was filed for a hand grenade type of device designed to put out fires. It took the form of a glass bottle, containing an often toxic liquid such as carbon tetrachloride. When thrown at a small fire, the bottle shattered and the liquid extinguished the flames.

As railway mania swept the Victorian world, it became clear that the way forward would be for steam trains running on two rails, but that didn't stop some going against the trend with less conventional types of railway. In England, for a while it seemed pneumatic railways might become popular, in which carriages were blown along air-tight tunnels by huge fans. In America, inventor Joseph Meigs came up with an elevated steam railway in which the engine and its carriages were balanced on a monorail that ran on pillars, in some places high above the ground.

A plaque in Westminster commemorates the inventor of the first traffic lights.

Meigs' elevated railway.

The hand grenade that extinguished small fires.

Let us not forget also spectacles for short-sighted horses, bicycles that ran on rails, electric-powered submarines, electric shock therapy, cameras disguised as guns, electric chicken hatchers, exploding scarecrows, hinged lamp posts and so much more.

THE INGENIOUS VICTORIANS

To belittle, or laugh at, some of these ideas is to misunderstand the Victorian urge to continually find astounding and sometimes bizarre ways of carrying out everyday, as well as hitherto impossible, tasks. Even their failures should be applauded for the effort put into trying something new or revolutionary. Many of these failures are fascinating in themselves, but of course they pale into insignificance beside some of the great and enormously successful inventions, concepts and achievements of the age.

It's these amazing achievements and, yes, some of the heroic failures too, that are celebrated on the following pages, in honour of those sometimes weird, often wonderful, always inventive, innovative and ingenious Victorians.

CHAPTER 1

The World's Biggest Glass Building

In 1847, Prince Albert, husband and consort of Queen Victoria, took a leading role in the organisation of a huge international exhibition of culture and industry from around the world. London's Leicester Square was initially considered for the exhibition site, but soon rejected as being too small, and plans were shifted to Hyde Park. A competition was launched, inviting British and European architects to submit designs for the building that would house the exhibition. Although 245 designs were received, the Building Committee didn't like any of them and set about producing their own, much to the disapproval of English gardener, designer, writer and successful speculator, Joseph Paxton.

To call Paxton a mere gardener is somewhat misleading. Born in 1803, Paxton's early working life was at Chiswick Gardens, parkland

Sir Joseph Paxton, as depicted in The Illustrated London News *in 1851.*

that was leased by the Horticultural Society from the Duke of Devonshire. In 1826, impressed by the young Paxton's abilities, the Duke appointed him Head Gardner at Chatsworth House, the Duke's country house in Derbyshire. In this position, he designed gardens and fountains, as well as a model village, arboretum and a special lily house whose design was based on the shape of the leaves of a giant species of the plant. More significantly, in view of what was to come, he built for the Duke a large glasshouse, which became known as the Great Conservatory.

While at Chatsworth House, Paxton was allowed the freedom to work on other projects, which included helping with improvements to the Royal Botanic Gardens at Kew, design of a country house for Baron Mayer de Rothschild and lending his expertise to the design and layout of various public parks.

He was also keenly interesting in all things modern and mechanical, which led him to successful investment in the boom that the railway industry was enjoying at that time. As well as designing a fashionable nineteenth-century estate of villas based around a circular road, he contributed to the designs of railway stations in the Midlands, and also built one of his own design. All of this helped to make him a rich man, and in 1848 he was appointed as a director of the Midland Railway. It was while presiding over a railway board meeting that the genesis of what became the building that would house the forthcoming exhibition began as no more than Paxton's doodling on a piece of blotting paper.

Paxton's design
When he heard that members of the exhibition's Building Committee had rejected all the designs submitted to them and decided to press on with plans of their own, Paxton had his doubts. The committee wanted bricks and mortar. Paxton, thinking back to the Great Conservatory that he had built for the Duke of Devonshire, thought more in terms of iron and glass.

The committee's plans were due to be announced in two weeks. Paxton told them he would have his plans completed and ready for presentation in nine days. Based on his first doodles on that piece of blotting paper, he engaged a small team, and together they worked day and night to complete the plans.

The design was revolutionary, but any doubts the committee might have had were swept aside when Paxton went over their heads and had

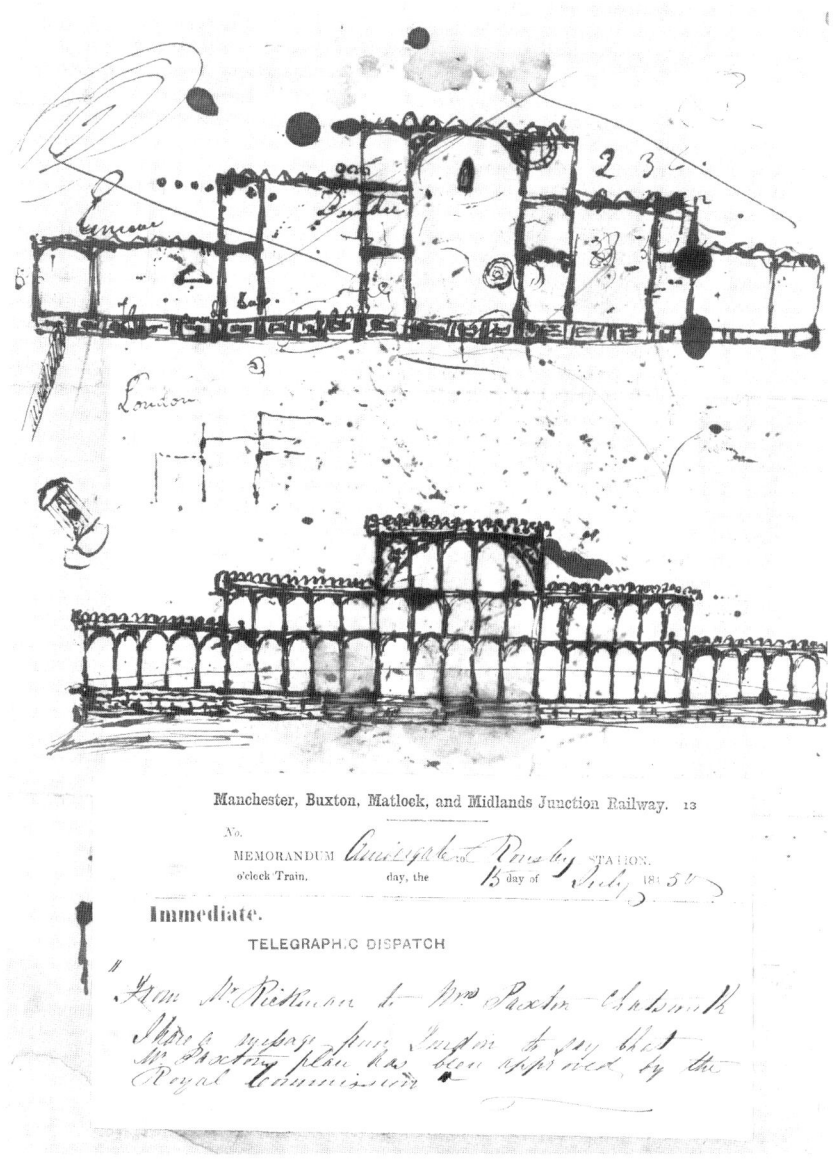

Paxton's initial doodle that was the starting point for the Crystal Palace.

the plans published in *The Illustrated London News*, where they won public acclaim. The committee gave in to the inevitable and accepted Paxton's plans, with at least one major modification. Three giant elm trees stood across the path of the building site. There had already been some opposition to the way the building might decimate areas of Hyde Park,

Work begins on the Crystal Palace.

Machinery used in the construction for drilling and punching.

Constructing the roof of the immense building.

Glaziers work from a mobile wagon as it runs on wheels along the guttering.

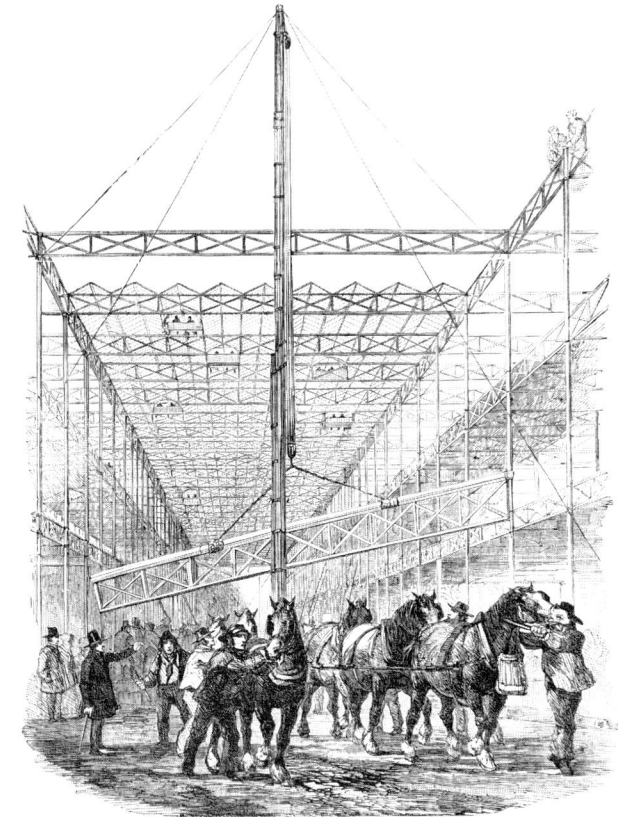

Horses were used to raise the girders.

Raising the ribs of the transept roof.

and fearing condemnation for felling the trees, the committee instructed Paxton to incorporate them in his plans.

Work began in July 1850, with thirty-nine workmen. By the time it was finished, more than 2,000 workmen had been engaged.

It began with stakes driven into the ground to indicate the position of the building's columns. Every casting delivered was examined and

THE WORLD'S BIGGEST GLASS BUILDING

painted before being incorporated into the structure. Only the choicest timber was selected. Up to 500 painters were employed. Teams of glaziers put in the glass, at the rate of about 18,000 panes a week.

Three times the length of St Paul's Cathedral, the Crystal Palace was the largest glass structure in the world. It stood on a plot 1,848 x 408 feet, plus an addition on the north side of 936 x 48 feet, with a height at its tallest point of 108 feet. It contained 330 huge iron columns, 24 miles of guttering, using 4,500 tons of iron and 600,000 cubic feet of timber. Fitting 293,655 panes of glass, glaziers sat in covered wagons whose wooden wheels ran along the bottom of the guttering. The wagons could be covered in tarpaulins so that work might proceed in bad weather.

Never before had anyone seen such a vast expanse of glass in a building. The interior was light and warm, needing little artificial illumination or heating during the day. There were three entrances, seventeen exits and ten double staircases to the galleries. The building was decorated in four colours. Inside, vertical flat surfaces were painted light blue, rounded parts were yellow and the undersides of girders were red. Outside, the exterior was white, with blue details.

South entrance of the Crystal Palace during The Great Exhibition of 1851.

THE INGENIOUS VICTORIANS

At the centre of the interior stood a huge fountain, 27 feet high and made from 4 tons of pink glass. Trees already on the site before construction began were incorporated into the structure, and continued to grow inside the building.

Several names were suggested for the building – The Glass Palace, The People's Palace – before *Punch* magazine dubbed it The Crystal Palace, and the name stuck.

From July 1850, when possession of the land had been taken, the building of the Crystal Palace was completed in just six months, finished by January 1851. The exhibition that the Crystal Palace was built to house was known as The Great Exhibition. From around the world, goods began arriving for it in February, and on 1 May 1851, the exhibition was opened by Queen Victoria. After much acclaim and great success, it closed a little more than five months later, when Joseph Paxton was knighted by the Queen in recognition of his work.

But as the exhibition closed, a new chapter was about to begin for the Crystal Palace.

CHAPTER 2

The Great Exhibition

In 1851, Britain was in the midst of a vast manufacturing boom. Inspired by Prince Albert, the Queen's consort, it was decided that the time was right to show off the country's expertise to the world, and for the world to come to Britain to show what other countries also had to offer and boast about. The official name for this mighty endeavour was The Great Exhibition of the Works of Industry of all Nations, but it soon became known simply as The Great Exhibition.

The Gathering of Nations, *by French engraver, Antoine Johannot.*

THE INGENIOUS VICTORIANS

To help design and organise the event, Sir Henry Cole was called in. He was an English inventor who, among other things, was credited with playing a key role, alongside Rowland Hill, in the introduction of the Penny Post, and also for introducing the world's first commercial Christmas card. Industrial design was a particular passion of Cole's and he had already designed a prize-winning teapot made by Minton, as well as writing a series of children's books. Also on board were members of the Royal Society for the Encouragement of Arts, Manufactures and Commerce.

The idea wasn't entirely original. It had been partly inspired by the French Industrial Exposition of 1844. The British answer, however, would be much bigger, bolder and more impressive, to make it clear to the world that Great Britain was an industrial leader. To establish viability of such a project the government set up the Royal Commission for the Exhibition of 1851. The founding President of the Commission was Prince Albert, and Sir Henry Cole was Chief Administrator. Within a month of the completion of the Crystal Palace in January 1851, exhibits began arriving from all around the world, to be grouped into four categories: manufacturers, raw materials, machinery and fine arts.

The official opening
The opening day had been planned right from the start to be Thursday, 1 May. With precision timing, everything was complete and ready by exactly that date, much to the relief, not to say surprise, of many of the promoters.

On the afternoon of the day before the official opening, two companies of the Grenadier Guards assisted police in clearing the Crystal Palace of every person not concerned with the opening, to allow those who were involved to make the necessary preparations. Here's how *The Illustrated London News* described the following morning:

> Never dawned a brighter morn than on this ever-memorable May day; the sky clear blue, the sun shining forth in undimmed splendour, the air crisp, cool, yet genial as a poet's spring morn should be. London, with her countless thousands, was early afoot. By six o'clock, the hour fixed for opening the park gates, streams of carriages, all filled with gaily-attired company, came pouring from all parts of the metropolis and the surrounding districts, while whole masses of pedestrians marched in mighty phalanx towards the scene of action. All St James's Park, all the way up

THE GREAT EXHIBITION

Constitution Hill, all the way along Knightsbridge and Rotten Row was now one sea of heads whose owners were intent on one object – to catch a glimpse of Her Majesty and splendid suite on the way to the Palace of Industry.

At eleven o'clock that morning, the Horse Guards began widening a path through the crowds, and at half past eleven, as the regiment's band began playing *God Save The Queen*, the royal party of eight carriages set out from Buckingham Palace, heading for Hyde Park. Fifteen minutes later, they arrived at the northern entrance to the Crystal Palace. As the Queen and her party entered the exhibition, the Royal Standard was unfurled from the top of the building.

Inside the Crystal Palace, at the place where sectors designated in church-like fashion as the naves and transept intercepted, a platform had been erected for the Queen and other dignitaries. A canopy trimmed with blue satin and draperies was suspended overhead, and before Her

Queen Victoria's arrival at the Crystal Palace.

Majesty's chair stood an enormous pink glass fountain. Flowers and banners proliferated all around.

There followed a rather long-winded declaration from the exhibition's administrators, a reply from the Queen and a prayer led by the Archbishop of Canterbury, followed by a performance by the choir of the *Hallelujah Chorus* from Handel's *Messiah*.

The doors had been open to the public from nine o'clock, but at eleven they were closed again. Over the course of the next hour, a great many dignitaries arrived until, at twelve o'clock, heralded by a fanfare of trumpets, the royal party arrived and the Queen took her seat as a choir of a thousand voices sang the National Anthem. She wore a rich embroidered pink satin dress set with precious stones, and a diamond tiara. Prince Albert wore a field marshal's uniform.

The transept and fountain at the centre of The Great Exhibition.

Inauguration by Queen Victoria as she officially opened the exhibition.

The royal party then rose and proceeded in procession to tour the aisles of the exhibition, as several organs from the British contingency played successively along the route of the procession. *The Illustrated London News* reported that 'Throughout the whole of the Queen's traverse of the building, her face was wreathed with smiles and pleasant looks.'

On her return to the platform, the Queen declared the exhibition opened, as a flourish of trumpets relayed the message to waiting crowds outside and a royal salute was fired from the north bank of Hyde Park's Serpentine lake.

Queen Victoria then returned to Buckingham Palace along the route by which she had arrived, and the doors of the Crystal Palace were thrown open again to admit the waiting crowds.

In her journal, the Queen later reported:

That we felt happy and thankful I need not say, proud of all that had passed and of my beloved's success. I was more impressed by the scene I had witnessed than words can say. Dearest Albert's name is forever immortalised.

THE INGENIOUS VICTORIANS

The exhibits

Part of the exhibition was divided into courts, each of which displayed products from the many countries taking part. As the host country, Britain had the largest area, taking up about half the available space to show off the height and, to some extent, the eccentricities of Victorian ingenuity.

Among the British displays was a church pulpit connected by rubber tubes to pews in order to carry the sound of a preacher's voice to hard-of-hearing members of the congregation. Another oddity on display was ink that produced raised characters on paper so that the blind could feel the letters.

 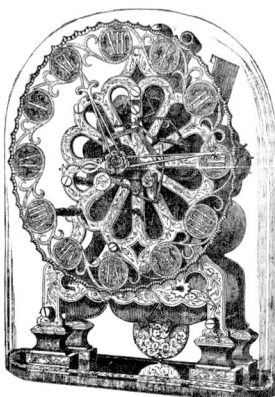

Two clocks from British clockmakers R. & J. Moore.

Dressing case, inkstands and caskets by British jewellers Asprey.

An enormous silk trophy made by British manufacturer Keith & Co.

Rather more practical was massive hydraulic machinery that had been used to lift tubes, each weighing 1,144 tons, into place for a bridge built at Bangor. Second only in size to that was an enormous steam hammer, so accurate that it could forge the main bearing of a ship, or just as easily and gently, crack an egg.

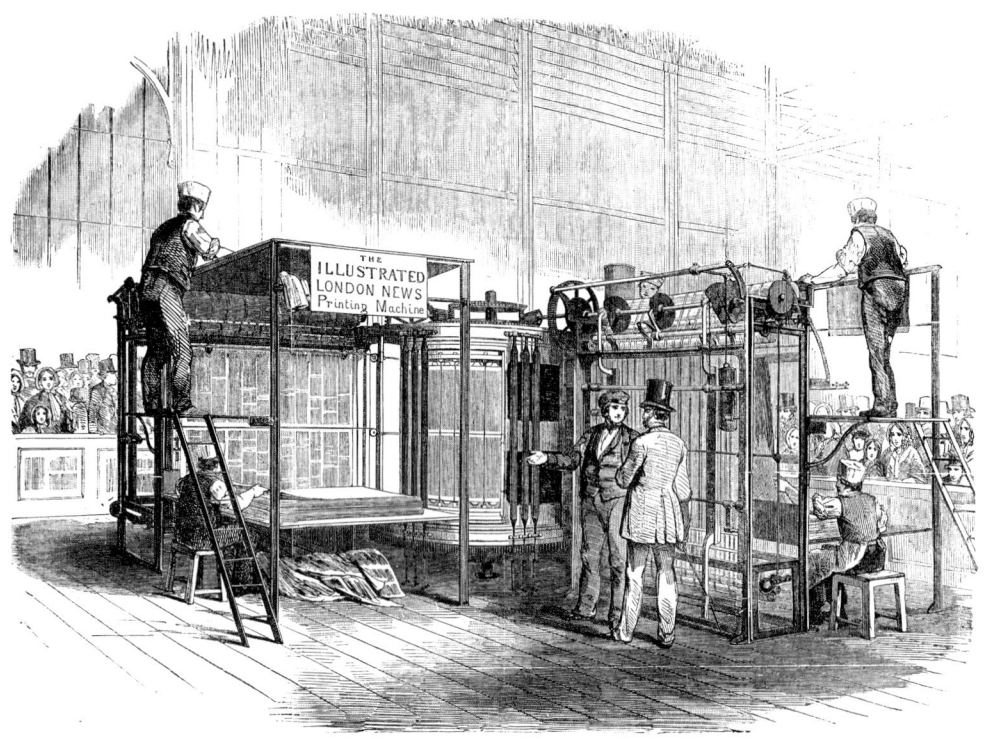

The Illustrated London News *being printed at the exhibition.*

One gallery was walled with stained glass to produce multicoloured patterns as the sun shone through it. There were carpets from Axminister, ribbons from Coventry and an eighty-blade knife from Sheffield.

One of the most popular periodicals of the time was *The Illustrated London News,* which printed and published pages of up-to-the-minute information from around the world, set in tiny, tightly composed type and illustrated by skilfully produced wood engravings. To produce this on a weekly basis was no mean feat, and at the exhibition, the publishers had on display and working their printing machine, which was capable of turning out 5,000 copies of the periodical per hour.

Another machine, for printing and folding envelopes, was also shown, along with a device for making cigarettes, the smoking of which was becoming increasingly popular. Other British-made marvels on show were examples of textile and agricultural machinery, as well as a range of different types of steam engine. One whole gallery was devoted to elegant carriages and a few examples of the velocipedes that predated bicycles.

An envelope-folding machine, displayed by British printers Delarue.

The Belgian Court.

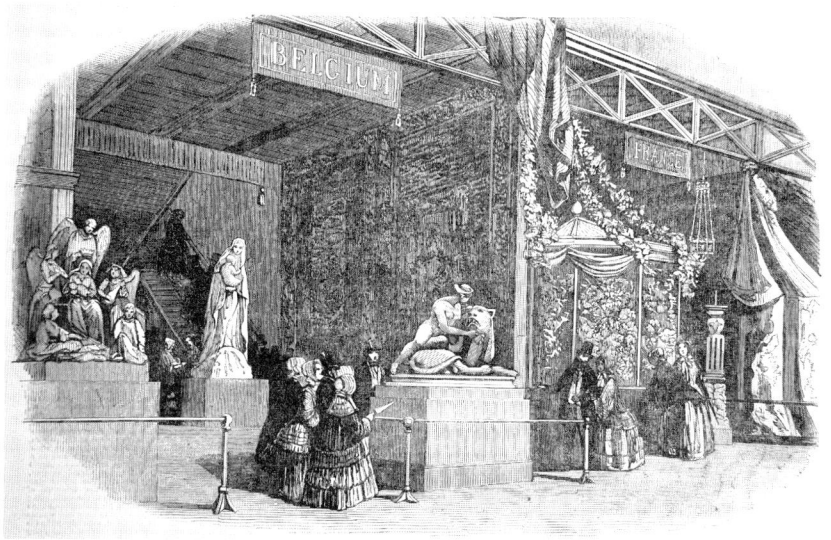

Less practical perhaps was the expanding hearse, not to mention a folding piano, aimed at yachtsmen who wanted a little portability in the enclosed space of their yachts.

After Britain, France was the largest contributor, producing a stunning display of silks, tapestries, porcelain and furniture, along with some of the machinery used to produce them.

The East Indian Court.

From America came a huge eagle whose outstretched wings held a large American flag. The Americans showed examples of firearms, even though the exhibition was supposed to be promoting world peace, but alongside these were also examples of the latest in reaping machine technology.

Among the Indian exhibits there were models of fishing boats, brass vessels used by the people of India, gold-embroidered shawls and a carved ivory throne. From Egypt and Tunis came matchlocks, sabres and large, magnificently embroidered saddles. Portugal showed a giant dripstone said to possess powerful filtering properties. Mosaics, marble, walnut and ornamental stones were displayed by Italy. Deadly swords and rapiers that contrasted with samples of raisins, figs, corn, wheat, tobacco and other agricultural productions were high spots of Spain's exhibits.

Among the exhibits from Turkey was a brazier, described as being like a huge tea urn, which burnt charcoal for the purpose of heating apartments. Canada sent a fire engine with panels painted with Canadian scenes. From Russia came huge porcelain vases and urns, furs, sledges and Cossack armour. A single lump of gold weighing nearly 8 stone arrived from Chile. Germany (at that time a collection of small states known as the German Customs Union) supplied small stuffed animals

The Tunis Court.

The Canadian Court.

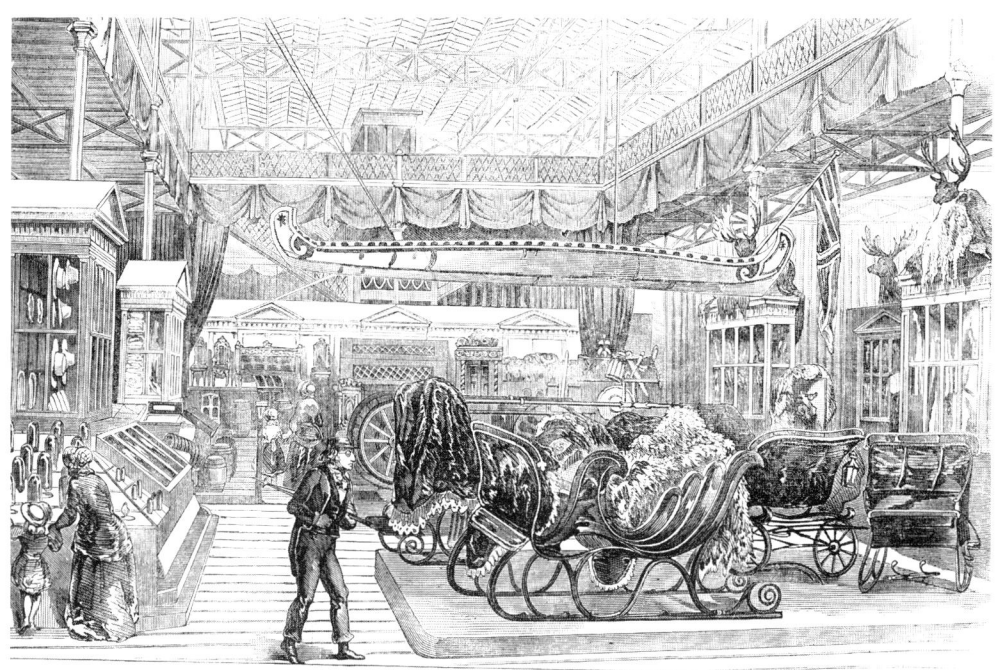

arranged in a series of charming tableaux that showed them in human-like activities. Switzerland, predictably, sent watches.

Also on show was the famous Koh-i-Noor diamond, once the largest known diamond in the world and later cut to be part of the British Crown Jewels.

The Ceylon Court.

The Russian Court.

THE GREAT EXHIBITION

To list only a small selection of the thousands of exhibits from just some of the many countries that contributed to the exhibition barely scratches the surface of all that was on show. In all, more than 100,000 objects were displayed along approximately 10 miles of floor space, by more than 15,000 contributors.

Charlotte Brontë, author of *Jane Eyre* and one of the exhibition's many celebrity visitors, summed it all up in *The Brontës' Life and Letters*, published in 1907, in which she wrote:

> It is a wonderful place – vast, strange, new and impossible to describe. Its grandeur does not consist in one thing, but in the unique assemblage of all things. Whatever human industry has created you find there, from the great compartments filled with railway engines and boilers, with mill machinery in full work, with splendid carriages of all kinds, with harness of every description, to the glass-covered and velvet-spread stands loaded with the most gorgeous work of the goldsmith and silversmith, and the carefully guarded caskets full of real diamonds and pearls worth hundreds of thousands of pounds. It seems as if only magic could have gathered this mass of wealth from all the ends of the earth – as if none but supernatural hands could have arranged it, with such a blaze and contrast of colours and marvellous power of effect.

On a more down-to-earth note, The Great Exhibition also saw the first major installation of public toilets, complete with flushing lavatories. Visitors to the exhibition were charged one penny to visit them – from which came the old euphemism of 'spending a penny'.

After the exhibition
The Great Exhibition closed on 11 October 1851. While it was open, it attracted more than 6 million visitors, even though Sunday opening was not allowed. Neither were alcohol, smoking or dogs. Initially, the price of admission was £3 for men and £2 for ladies. Later, the masses were let in for one shilling (5p) per head. The exhibition made a profit of £186,000, most of which went towards building the museums that now stand in South Kensington.

The Albert Memorial, one of London's most ornate monuments standing in Kensington Gardens, is often thought to have been erected in honour of Prince Albert's work in bringing The Great Exhibition into being, especially since the Prince is depicted, sitting under a canopy and holding a copy of the exhibition catalogue against his knee. In fact, the Albert Memorial was commissioned and erected more as a general tribute to the Prince's memory.

The real memorial to The Great Exhibition takes the form of a more modest, but nevertheless still impressive, statue to be found across the road. Commissioned from sculptor Joseph Durham, the original intention was to commemorate the exhibition with a figure of Britannia as the centrepiece of the monument. But after Prince Albert's death in 1861, the decision was made to put Albert in place of Britannia. The monument was first erected in the gardens of the Royal Horticultural Society at Kensington in 1863. In 1890, it was moved to its new site behind the Royal Albert Hall, where it remains today as a lasting tribute to Prince Albert and his amazing vision of The Great Exhibition.

The memorial to The Great Exhibition and Prince Albert's participation in it stands today behind the Royal Albert Hall.

CHAPTER 3

The Crystal Palace Reborn

Following the closure of The Great Exhibition in October 1851, the problem of what then to do with the Crystal Palace became apparent. It had been agreed from the start that, once the exhibition was over, Hyde Park would be returned to its original status. So what to do with the world's largest glass building? Once again, the newly knighted Sir Joseph Paxton came to the rescue, with a bold and ambitions plan to tear down the Crystal Palace and rebuild it on the other side of London, where it would form the centrepiece of a park full of wonders.

Under his direction, the huge iron and glass building was taken apart

Building the Egyptian Court.

Colossal figures from Abu Simbel inside the South London Crystal Palace.

THE CRYSTAL PALACE REBORN

and transported piece by piece to the top of Sydenham Hill in South London. Here, it was rebuilt, in a style that emulated the original, without being an exact copy, and also larger and higher. Work was completed in 1854, and the result was even more impressive than it had been in its previous incarnation.

Inside, the Palace became a museum of world cultures, depicting both ancient and modern civilizations. Its central transept, with a diameter twice that of the dome of St Paul's Cathedral, could seat up to 4,000 people. Its organ, with air supplied by hydraulic machinery, boasted 4,384 pipes. The transept was used for numerous events and music concerts.

Outside, the surrounding landscape of terraces, statues, ornamental English and Italian style gardens, gave Victorians a glimpse of what a future theme park might look like.

At each side of the Palace, huge water towers were built to provide the volume and pressure of water required to power fountains that were said to rival those at the French Palace of Versailles. Containing nearly 12,000 jets and requiring more than 7 million gallons of water for one display, they were soon deemed too expensive to run and were rarely seen in action.

More successful attractions included tidal lakes, with rocks and rapids, where visitors could take out boats or plunge from a great height down a water chute; the country's largest maze; an area devoted to the study of geology; and the frightening Topsy-Turvy Railway. A forerunner

The fountains of the new Crystal Palace when it was rebuilt in South London.

The boating lake and water chute in the first Victorian theme park.

The frightening Topsy-Turvy Railway, like a Victorian roller coaster.

THE CRYSTAL PALACE REBORN

A souvenir medal issued for visitors to the Crystal Palace Park.

of today's roller coasters, this involved cars that took two or four passengers, which raced down a steep incline, building up enough speed to completely loop the loop as the track turned in a vertical circle.

Another area of the park exhibited life-size replicas of dinosaurs and other creatures, displayed on islands in a lake that represented a timeline through three prehistoric eras. As the tidal water rose and fell, different aspects of the dinosaurs were revealed.

The Victorian prehistoric creatures have survived and been refurbished.

The Chinese Pavilion at the 1911 Festival of Empire.

The parkland was also the venue for twenty FA Cup Finals from 1895 to 1914, during which, in 1905, Crystal Palace Football Club was formed.

In the years that followed The Great Exhibition, the Crystal Palace became the venue for concerts and shows whose subjects included flowers, pigeons, poultry, goats, rabbits, cats, dogs, cattle and birds; and exhibitions of electrical items, art, aeronautics, mining, photography and transport. Societies that ranged from the National Temperance League to German gymnasts met and held rallies there.

Charles Blondin, the man famous for walking a tightrope across Niagara Falls in 1859, made an appearance where, posters proclaimed, he would perform Unrivalled Feats, which included walking, running and cycling along a tightrope, and cooking an omelet in mid-air.

The Palace was particularly famous for its Thursday night firework displays. They included such delights as a re-enactment of the bombardment of Alexandria in 1882, the flight of twenty monster dragonflies, immense cascades of fire and a finale of 2,000 rockets. Dr Lynn, the electrifying conjurer, also made an appearance.

Decline of the Palace
After Queen Victoria's death, as the Victorian age turned into the

THE CRYSTAL PALACE REBORN

Edwardian era, the Crystal Palace continued to enjoy a limited success. In 1911, the Palace played host to the Festival of Empire, in which the countries of the British Empire exhibited products in specially built scale models of their parliamentary buildings. The festival also included a pageant with tableaux illustrating the history of London and the Empire.

It was among the last of the great events at the venue, and despite attracting visitors from across the world, it still did not produce enough revenue to resolve the financial problems that the Palace had been suffering for some years previous – a situation that eventually saw the Crystal Palace go into liquidation.

On the evening of Monday, 30 November 1936, people living in and around the area noticed a red glow in the sky coming from the direction of the Crystal Palace. A fire had started in the office area of the centre transept. Despite the efforts of night watchmen to put it out, the fire quickly spread, and soon engulfed the whole building.

The first fire engine from Penge Fire Station arrived just after 8.00 pm. During the night, eighty-eight fire engines and more than 400 firemen fought a losing battle against the blaze, the flames of which could be seen from miles around.

In the morning, the wonderful and amazing Crystal Palace was no more. Arson was suspected, but never proved, due to the size of the building and the immense amount of flammable material that it contained. For many years, only the dinosaurs and the maze survived. Otherwise, the remains of what was once the site of this fantastic building were little more than overgrown terraces, flights of steps leading nowhere and damaged statues.

Steps up to the terraces are part of the little that remains of the great building today.

THE INGENIOUS VICTORIANS

One of two existing sphinxes that guard a staircase to what was once an entrance to the Crystal Palace.

A postscript to the Palace

One interesting postscript to the Crystal Palace story is the way it, and the exhibition it was originally created to house, gave other countries the impetus to try something similar, and so began a series of trade exhibitions across the world. The first of these – The Exhibition of the Industry of All Nations – opened in New York in 1853, only two years after London's Great Exhibition, and to house it, the Americans built their own version of the Crystal Palace.

As with the London event, the New Yorkers held a competition to design a building for their exhibition, and Sir Joseph Paxton was among the unsuccessful entrants. The competition was won by Georg Carstensen, a Danish army officer who had been one of the developers of the Tivoli Gardens, a famous amusement park and pleasure garden in Copenhagen.

Although by no means a copy of the original, the American Crystal

THE CRYSTAL PALACE REBORN

Palace did bear a striking resemblance to the London building, apart from a huge dome, 100 feet in diameter, which rose from, and dominated, the centre of the iron and glass structure.

Five years after its opening in October 1853, while it was housing the Annual Fair of the American Institute, the New York Crystal Palace caught fire, and within twenty-five minutes burned to the ground, a strange portent of what would befall the London Crystal Palace nearly eighty years later.

A huge bust of Sir Joseph Paxton remains today and towers over Crystal Palace Park.

CHAPTER 4

The Great Victorian Way

Even as the reincarnation of the Crystal Palace opened for business in South London in 1854, Sir Joseph Paxton was busy formulating a new project that would be even more impressive. His plan was for another enormous glass-roofed construction, but this time not merely a single building. His plan was for a huge 10-mile long tunnel, with a design obviously influenced by the Crystal Palace, that would encircle the whole of the centre of London.

It would comprise a central boulevard where people could stroll or drive horse-drawn carriages and other vehicles. At each side of the boulevard there would be high-class shops and superior living accommodation; it would pass through areas for recreation; and a two-tier railway was planned to run around the perimeter. Built with a domed roof of glass panels, it would be light, warm and comfortable to walk, ride, shop and live in, free of the cold, damp, rain, fog and smoke of the grimy outside world. It was to be called the Great Victorian Way.

Paxton's plans for this amazingly ambitious project were drawn up at the request of the government and submitted to the Parliamentary Select Committee on Metropolitan Communications in 1855. Its purpose was threefold. First, it would help to rationalise the crooked nature of many of London's streets. Second, since it was designed to link London's major railway station termini, it would cut out the road congestion, which was a major problem in Victorian London, and speed up journeys between any of those stations. Third, the route planned would link some of the poorest areas of the capital with some of the richest, and help towards the regeneration of slum districts. It was thought that people living in the poorer areas of London might be given a route to the more affluent districts of the capital, helping them aspire more to the betterment of their lives.

Here's the route that The Way was planned to take. Starting at

THE GREAT VICTORIAN WAY

Cheapside in the City of London, it travelled to Mansion House, across Cannon Street, and crossed the river Thames between Southwark and Blackfriars bridges, following the line of the Cannon Street Railway. From there, it went along the south of the river to Lambeth, then crossed back again over the Thames south of the Houses of Parliament. Passing Victoria Street, then through Sloan Square, it took a route through Brompton and Kensington Gardens before arriving at Paddington Station's Great Western Railway terminus. From there, it cut through Marylebone, then moved east between Oxford Street and Regent's Park to reach the terminus of the London and North Western Railway at Euston and the Great Northern Railway terminus at King's Cross. Moving then through South Islington and Aldersgate, it arrived back at its starting point at Cheapside.

Also planned was a branch line that ran inside the circle, from Lambeth to the South Western Railway's terminus at Waterloo, before crossing the Thames yet again between Hungerford and Waterloo bridges, and journeying onward to Piccadilly Circus.

A cross section of the Great Victorian Way, as envisaged in Paxton's watercolour painting.

THE INGENIOUS VICTORIANS

Inside The Way
The structure of the Great Victorian Way was very obviously influenced by Paxton's designs for the Crystal Palace. At 72 feet, its width was the same as that of the central transept of the Palace, and its height was 108 feet. The walls were to be covered in ceramic tiles; the roof was domed and made from hundreds of sheets of plate glass. The overall design was typified by elegant pillars, arches and walkways on three levels.

The shops were planned to line each side of the boulevard between the City of London and Regent Street. As the structure moved through Brompton and into parts of West London, the shops would give way to the living accommodation on four floors. As it passed through Kensington Gardens, there would be wide open spaces for recreation and the encouragement of exercise. There would be a good number of libraries and reading rooms along its route.

Being under cover all the way, it was envisaged that The Way would remain dry, and without rain there would be no mud, as was the case on the outside roads. The basic design and shape of the structure would remain much the same whether it travelled across land or over the river, in effect making bridges on which people might both shop and live.

The railway would run behind the shops and private residences, using two levels, one for fast trains and one for slow, with a double wall between the railway lines and any living accommodation to cut down on noise. The possibility of noise would also be reduced because this was an atmospheric railway.

The basic principle of the type of atmospheric railway to be used in the Great Victorian Way used air pressure to drive a bullet-shaped piston along a metal tube laid between and beneath the rails, in which a vacuum was created by pumping devices at the side of the line. The piston was attached, by a bar and via air-tight leather flaps, to the underside of the train. As the piston moved along the tube, it dragged the carriages along the tracks above. The Victorians, in search of a technology both cheaper and cleaner than steam, had been experimenting with atmospheric railways for a decade or more, and so far had had only limited successes when built along traditional railway routes. Paxton was convinced, however, that within the controlled conditions inside his structure – a place where steam would not be entirely impractical, but certainly not eminently desirable – an atmospheric railway stood a much better chance of success. Where outside streets crossed The Way, the railways would run above.

Because the structure linked all the major railway termini of London, it could be used to quickly transport goods from one terminus to another, although there was no direct link to existing steam railway systems. It was estimated that the railways would carry more than 100,000 passengers a day.

Within The Way there were to be strict operating times for various types of vehicle. Before nine o'clock in the morning, various tradespeople, such as coal men and the like, would be allowed to deliver goods to shops and housing accommodation. After nine o'clock, the only vehicles allowed in would be passenger carriages and omnibuses.

Paxton's rivals
Paxton's Great Victorian Way was not the first attempt at producing something similar. A short while before Paxton presented his plans, the government had considered a project from London architect William Mosely. His idea was to be called the Crystal Way and was very similar to, though less opulent than, the Great Victorian Way. Not long after, Frederick Gye, Director of Covent Garden's Royal Italian Opera, claimed that he had put forward an idea for a Glass Street some years before in a paper delivered to the Institute of Civil Engineers. Ideas that used iron rather than glass had also been suggested, including one called the Super-Way, which was proposed around the same time as Paxton's Way by Parliamentarian John Pym.

But it was Paxton, still basking in the glory of the Crystal Palace, who had his ideas taken seriously. On 7 June 1855, he made his proposal, endorsed by Prince Albert, to the Select Committee, using an aerial drawing and a detailed watercolour, showing a cross section of The Way. Paxton was meticulously questioned by the eight-man committee, chaired by William Jackson, which led to an Act of Parliament being passed to authorise its construction.

Such an ambitious venture would obviously not come cheap. Paxton estimated that it would cost £34 million, an enormous amount of money in the mid-nineteenth century and one that eventually led to the scheme's downfall. Despite Paxton's assurances that much of the cost might be hopefully offset by income from the shops and living accommodation, the government ultimately decided that the funds would be better spent on the creation of a functioning sewer system for London, and plans for the Great Victorian Way were abandoned.

Some time later, with the building of the London Underground, the Circle Line was built to roughly follow, under the ground, the same route as The Way. While an underground railway line could obviously not provide the luxuries of shops, homes, libraries and a central boulevard for strolling, it did nevertheless address many of the transport benefits that Paxton had originally envisaged for the Great Victorian Way.

CHAPTER 5

When *Daddy Long-Legs* Ruled the Waves

In 1883, an innovative inventor named Magnus Volk opened Britain's first electric railway at Brighton in Sussex. It proved to be a successful tourist attraction, and its successor still runs today, the oldest electric railway in the world. It was when Volk decided, in 1890, to extend the line east to Rottingdean, that things took a turn for the bizarre.

The new line would have to scale a steep incline to the top of a cliff, or be built into the unstable undercliff, and neither was practical. So with Victorian resourcefulness, Volk came up with an unusual plan. He would lay the railway lines under the sea and run the train on long steel legs 24 feet above the waves. He called it the Brighton and Rottingdean Seashore Electric Railway.

Magnus Volk, whose Germanic name came from being the son of a German clockmaker who had settled in Brighton, was a forthright, precise and fastidious person who, from an early age, was fascinated by electricity and model making. A master electrical engineer, Volk was the first in Brighton to own a telephone and to equip his house with electric light. Appointed as electrical engineer to the Corporation of Brighton in 1883, he soon installed electricity in the Brighton Pavilion, the town's one-time royal residence.

A few years later, Volk built a very early type of electric car, which he demonstrated on the sea front at Brighton. News of the car reached a German newspaper, where it was spotted by the Sultan of Turkey, who

Magnus Volk, creator of the Seashore Electric Railway.

contacted the inventor with a request to make a similar car for him. Volk not only built the car, but also delivered it in person, and while he was in Turkey, equipped the Sultan's palace with electricity.

Building the railway
Construction on the Seashore Electric Railway began in 1894, with labourers working only at low tide during a bitter winter that caused the sea water to freeze on the beach. Their job was to lay four rails, consisting of two parallel sets of two-rail tracks. The gauge of each pair was the standard railway gauge of 2 feet 8½ inches, and the distance between the outer rails of the four was 18 feet. The route was kept as level as possible, its steepest gradient no more than 1 in 300.

Square holes were dug in the chalk of the shore between the low and high tide marks and filled with concrete, boxed above the surface so that, when set, part of each embedded block stood above the ground. The tracks were then fastened to the blocks. When the tide came in, the sea covered the lines and the workmen retired until the next low tide.

A model of Volk's Pioneer *carriage.*

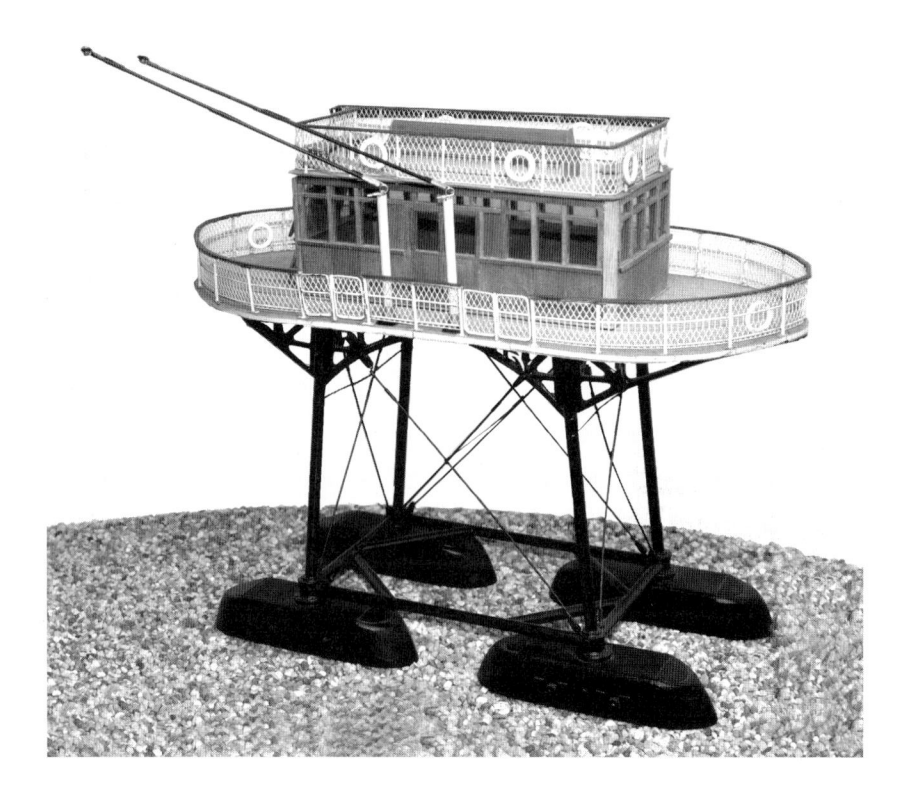

WHEN *DADDY LONG-LEGS* RULED THE WAVES

The single railway carriage was built by the Gloucester Railway Carriage and Wagon Company, which set up business at Gloucester in 1860. Among its products, the company made goods wagons, passenger coaches and stock for the London Underground. They also undertook special-purpose work – and Volk's project certainly came into that category.

The single carriage was like a tramcar crossed with a yacht on legs. Electricity to power the motors was generated by a gas-driven generator, producing 500 volts and housed below the pier at Rottingdean. It was supplied via twin connecting rods from the carriage, each tipped with small wheels to run along an overhead cable slung from poles beside the tracks. On board, 25-horsepower electric motors drove long shafts that ran inside the tubular legs, geared to the wheels, which were enclosed in metal casings, shaped to push debris out of their path as the contraption made its way through the sea. The wheels on the two legs on one side of the carriage ran on one set of tracks, and the wheels on the two legs on the other side ran on the second set of tracks. Volk called the carriage *Pioneer*. The public looked at its strange shape and renamed it *Daddy Long-Legs*.

An artist's impression of Daddy Long-Legs *in action, painted from a postcard of the time.*

How a voyage on Daddy Long-Legs *was advertised on posters of the time.*

    When the tide was out, the rails could be clearly seen. At high tide, they were submerged up to 15 feet below the surface.

    The electric current picked up from the overhead cable was returned to earth via the rails and thence into the sea. Although it seems there was no immediate danger, it has to be said that the mixture of electricity and water is never a good one and must have been unnerving for bathers and paddlers in and around the passing carriage.

    Although designed to run on rails and therefore justifiably called a railway carriage, *Pioneer* was actually classified as a seagoing vessel. As

WHEN *DADDY LONG-LEGS* RULED THE WAVES

such, maritime law required it to carry a lifeboat and lifebuoys, and the driver had to be a qualified sea captain.

Passengers alighted and disembarked at piers built out into the sea at Brighton and Rottingdean, with an intermediary stop-off pier at Ovingdean. Use of the piers put passengers on the same level as the carriage, high above the water level. Once aboard, they found themselves on a veranda 45 feet long and 22 feet wide that surrounded an enclosed saloon with an open promenade deck on its roof. The saloon was equipped with plush, upholstered seats, curtains and even plants. The carriage took a maximum of 160 passengers, and the whole thing weighed in the region of 45 tons.

The maiden voyage

Daddy Long-Legs made its first journey – or maiden voyage – on 28 November 1896. The mayor of Brighton, the chairman of Rottingdean Parish Council, the town's two Members of Parliament and other dignitaries were on board. Proceeding at a leisurely pace, the carriage's 2¾-mile journey took thirty-five minutes and, on its return, a celebratory luncheon took place.

Passengers boarding the carriage at Kemptown in Brighton.

The carriage running dry at low tide.

Daddy Long-Legs, *running through the water at high tide.*

WHEN *DADDY LONG-LEGS* RULED THE WAVES

Newspaper reports were enthusiastic. One suggested that *Pioneer* would revolutionise sea travel as trains once revolutionised land travel, envisaging a day when such a railway would run from Dover to Calais, or even London to New York, though how railway lines could be laid along the Atlantic seabed wasn't mentioned.

A week after *Pioneer's* launch, tragedy struck. It came during one of the worst storms ever remembered on this part of the south coast. *Pioneer*, lashed to its moorings at Rottingdean Pier, broke free to be overturned and smashed by the storm winds. Some of the rails and poles that held the overhead cables were also damaged.

Remarkably, Volk managed to salvage and rebuild *Pioneer* in time for a summer re-opening in July the following year. After that it ran for another three years, carrying thousands of passengers, including the then Prince of Wales, who made the trip twice on 20 February 1898. A ticket cost sixpence.

English novelist Angela Thirkell, in *Three Houses*, her book of childhood memories of the area in and around Rottingdean, later referred to it as 'the most preposterous machine … more like a vision of the Martians than anything else you ought to see at a peaceful seaside village'. Her comparison was with the long-legged Martian fighting machines described by H.G. Wells in *The War of the Worlds*, published in 1898, at the time *Pioneer* was running. Thirkell wrote:

> We were never allowed to go in it, partly because no grown-up thought it amusing enough to go with us and partly because it had a habit of sticking somewhere opposite the ventilation shaft of the Brighton main sewer and not being moved till nightfall. When it discharged its passengers at the pier it took on a fresh load and stalked back to Brighton, leaving us gaping in admiration.

The end of the line

The year 1900 finally sounded the death knell for *Daddy Long-Legs*. First, it had to be closed during the peak summer season while repairs were made to the concrete blocks that had been undermined. No sooner had that difficulty been overcome than the local council announced that sea defence work meant the *Pioneer's* rails would have to be moved further out to sea.

Daddy Long-Legs, *left to rot on the beach at Rottingdean and to be broken up for scrap.*

WHEN *DADDY LONG-LEGS* RULED THE WAVES

Cleary this was impractical. Even as it stood, when the tide was high and the water deep, *Pioneer's* speed was reduced to little more than walking pace. In fact, there were times, if the water was especially high, when it was impossible for the carriage to move much further than a few hundred yards from the pier before giving up and having to return. Moving the rails further out into deeper water, even if it had been possible to lay them, would have slowed *Pioneer* down more, and it wasn't practical to replace the electric motors with more powerful versions.

In 1901, council workmen moved in to remove some of the rails to carry out the projected work on sea defences. *Pioneer* was taken to Rottingdean Pier, where it remained for the following nine years, slowly rotting away, until it was broken up for scrap in 1910. The pier survived for about another four years, before also being demolished.

Today, the only sign of Volk's achievements are a few rows of jagged concrete blocks straggling along the shoreline between Brighton and Rottingdean. Few who even notice them, as the tide recedes to uncover their remains, realise that they mark the foundations for the tracks on which, for a few brief Victorian years, *Daddy Long-Legs* ruled the waves.

All that remains today: stone blocks at Rottingdean, part of the foundations on which the tracks were laid.

CHAPTER 6

James Wyld's Monster Globe

London's Leicester Square, in Victorian times, was not the most salubrious of places. It was neither respectable nor hygienic. In *Household Words*, a weekly magazine edited by author Charles Dickens, the Square was once described as 'a howling desert with broken railings, a receptacle for dead cats and every kind of abomination'. The area might have become more respectable if an early scheme for using it as the site for The Great Exhibition had come to fruition, but the idea was abandoned because the location was too small. So Leicester Square remained overgrown, dirty and a meeting place for ruffians and scoundrels. But then along came James Wyld with an unusual plan to change all that.

A former Liberal Member of Parliament for Bodmin in Cornwall, Wyld was a distinguished and prolific map-maker. His reputation was best summed up by *Punch* magazine, saying that if a country were discovered in the centre of the earth, Wyld would produce a map of it as soon as it was discovered, if not before.

James Wyld, Member of Parliament and map-maker who breathed new life into London's Leicester Square.

The plans
Wyld's plan began as a promotional project for his map-making business. His thinking was based on the fact that, from a single viewpoint, it would be impossible to see the whole of the Earth represented on the outside of a gigantic sphere. But if the same countries, continents, seas and oceans were represented on the inside of the sphere, and if people could go inside and look around, then the whole

60

world could be taken in from a single viewpoint – or at least several easily accessible viewpoints.

His plan was to build a huge globe, more than 60 feet in diameter. Inside that, staircases and elevated platforms would allow visitors to climb and view the surface of the earth on the concave inner surface, with the world's land masses, rivers, valleys and mountains built to scale from plaster of Paris. To quote what *Punch* said about it at the time, it was 'a geographical globule which the mind can take in at one swallow'.

Wyld originally wanted to site his globe in the Crystal Palace as part of The Great Exhibition. That, however, was rejected on the grounds that the Globe was too big, and also because the organisers were against any form of commercialism by exhibitors who attempted to use the exhibition to sell their goods or services. So he turned his attention to the site rejected by the exhibition organisers and set about acquiring land in Leicester Square.

Building the Globe
It wasn't easy. He received immediate opposition from many of the owners of buildings surrounding the Square. But after protracted negotiations with his adversaries, he eventually secured the land for £3,000 and building began. *The Illustrated London News* of 22 March 1851 stated:

> The dull and dreary centre of Leicester Square has of late become a scene of great activity in the commencement of the building to receive Mr Wyld's large Model of the Earth and surface. Upwards of 100 workmen are busily engaged upon this area. Huge balks of timber have displaced the stunted trees and shrubs; and large trusses have been framed for the support of the capacious structure.
>
> During the period of erecting and placing the trusses to support the dome in their several beds, those fixed have much the appearance of the ribs of a large ship. We congratulate the inhabitants of Leicester Square upon the prospect of having a building which, if the design is fully carried out, will be an ornament to the metropolis.

Within two weeks of the start of building, the workforce numbered about 300, working by day and by gaslight at night. They were constructing a

Building the Globe by gaslight.

circular building with four entrances and a diameter of 88 feet. Inside, thirty-two trusses supported the gigantic globe, which emerged as a huge dome from the roof, whilst vertical timbers inside supported four galleries. Everything was grounded solidly in concrete as it was thought that the slightest vibration might harm the delicate plaster of Paris modelling inside the globe.

During the building, a statue of George I on horseback had to be removed. To appease opposition to this, it was stated that the statue would be placed in a pit below the building and reinstated when the Globe was removed. Despite these assurances, rumours abounded that it had been sawn up and buried or that parts of it had been dismantled, stolen and sold by some of the workers. When later questioned on the subject, Wyld reported that the statue had been found to be no more than lead stuffed with clay, and that it was thrown away with other rubbish.

During the building, various unforeseen problems and setbacks involved new thinking to the design and construction, as well as an ensuing need for more money. The final cost was somewhere around £5,500, more than double the £2,000 that Wyld had planned for, but a lot less than an estimated £13,000 that would have been needed to complete the project if the construction of originally envisaged outlying buildings had not been cancelled.

One particular difficulty of construction concerned the plaster casts of the Earth's features. There were about 6,000 of these in all, each about 3 feet square. Each piece needed to butt up accurately against its adjacent piece whilst also maintaining a slight curvature so that it would perfectly fit against the concave inner shell of the Globe. It took three attempts to perfect the design.

Open to the public
Wyld's Great Globe – or Wyld's Monster Globe, as it was also known – was opened to the press and special guests on 29 May 1851, and on 2 June, it was opened to the public. It soon began to attract criticism from some quarters. Its critics thought the building's style of architecture was hideous and out of place against its surroundings. The residents of the Square, with whom Wyld had conducted somewhat delicate negotiations when he acquired the land, had expected a Crystal Palace type of glass structure and were outraged to find they were looking at a much uglier bricks and mortar building. The pubic, however, loved it, and it's not difficult to understand why.

From the outside: the building that housed the giant Globe.

THE INGENIOUS VICTORIANS

The four entrances, which faced north, south, east and west, took visitors into four large reception areas from where turnstiles led to a corridor that ran around the inner circumference of the building. The curved wall on one side of the corridor was the outside surface of the globe, painted blue with star constellations in silver. The opposite wall was hung with maps that promoted Wyld's business, and the corridor was lit by globe-shaped gas lamps. The pillars that supported the balconies were decorated with designs copied from the Alhambra Palace in Spain.

From this corridor, the public entered the interior of the Globe through an impressive entrance gate, decorated with Moorish patterns, influenced by Arabic designs.

On entering, visitors found themselves in the middle of the Pacific Ocean. Then, by means of platforms and staircases, they could wander the surface of the Earth, examining various areas of the world in close-up, the platforms taking them in places as close as 3 feet from the inner surface. Around the surface, the scale was 10 miles to the inch, but

From the inside: visitors view the surface of the earth from a series of staircases and walkways.

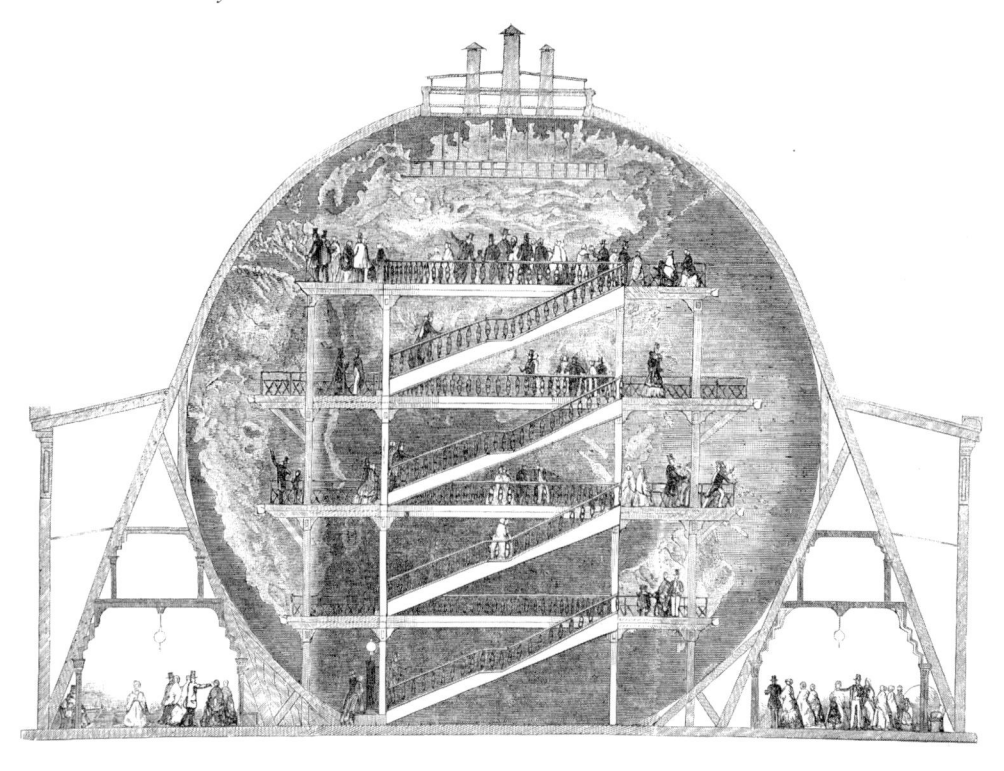

vertically, to give more drama to mountains ranges and craters, the scale was only 1 mile to the inch.

By day, the interior was lit by windows at the North Pole, which also provided a degree of ventilation. By night, the Globe was illuminated by gaslight. It was reported at the time that the combination of gas lighting, the heat of the human bodies and the minimum of ventilation resulted in extremely hot conditions. Since heat rises, the conditions were, of course, hottest at the North Pole.

One area not covered was Antarctica, at the South Pole, a continent that, at that time, was largely undiscovered and even thought by some to be non-existent. Other than that, visitors were often most impressed, and surprised, at the amount of water in the world, as opposed to dry land. *The Illustrated London News* of 7 June 1851, reported:

> The immense expanse of water in the southern hemisphere is brought out in strong contrast with the wide-spread lands of the northern; and the great chains of mountains which are remarkable features of the Earth's surface are shown to be ranged in a circle round the oceans and Indian Sea. The water-shed or river-course of every country is laid down and the great areas drained are exhibited.
>
> Mistakes may henceforth be avoided by the advantage of being now able to embrace at one view the whole of the Globe. Much is still to be learned by this means. The Globe in Leicester Square therefore will require to be constantly visited and will repay the student for renewed investigation.

From the day it opened, the Goble was a great success with the public, and was visited by Prince Albert, to whom Wyld had dedicated the project. It attracted thousands of visitors, particularly during the period that The Great Exhibition was bringing visitors to London from across the world, from May to October.

Being the showman and self-promoter that he was, Wyld produced a book called *Notes to Accompany Mr Wyld's Model of the Earth*. Used as a guide to the Globe, as well as a sales catalogue for his maps, the book provided a wealth of information, both geographic and historic, on London, Britain in general and the Victorian Empire. Wyld also gave regular lectures on the Earth, which he later expanded, when visitors began to decline, to include other popular themes of the day.

THE INGENIOUS VICTORIANS

Writing in the 12 July 1851 edition of Dickens's *Household Words*, Henry Morley, a professor of English Literature, commented:

> We walk about the surface of our globe, tread the hot flagstones of its towns, or crush the soft grass of its forests, bathe on the margin of its seas, float on its rivers, look abroad from its mountain-tops, and, like good common-place folk, here we say we are in town, there by the sea-side, there we are in the country. We walk into Leicester Square, and enter a neatly made brick packing case, look at the world boxed up in a diameter of 60 feet, and say, Ah, here is a colossal Globe! Here is a work of beauty! What a clever man its maker, Mr Wyld, must be!

With the closing of The Great Exhibition in October 1851, and the subsequent drop in people coming to London, interest in the Globe began to wane, although it continued to attract a significant number of visitors for some time.

The final days

As the months went on, however, interest diminished even more. Wyld formulated plans to sell the land and build a museum close to the site, but that never happened. There were complications concerning Wyld's right to sell, plus further problems with selling the building to a buyer who failed to pay. Court cases ensued, plus further legal wrangles that involved not removing the Globe from its location at a time previously agreed. Wyld eventually sold the edifice for demolition in October 1862.

By November that year, it was gone, and despite previously made promises to restore the Leicester Square Gardens, the area soon reverted to its former state of dilapidation. When the statue of George I was eventually unearthed, it was discovered that the rider had vanished – probably stolen and sold for scrap – and the horse's body was found lying on its side, with the head reportedly to have been found wrapped in sacking.

Undoubtedly, Wyld's Globe had been a major attraction and a great success in the limited time it was open, drawing in sizable numbers of visitors. But by the end of 1862, there was nothing left of what had once been one of the most remarkable, and certainly one of the most innovative, structures to have graced Victorian London.

CHAPTER 7

The Biggest Camera in the World

The Victorian talent for bizarre innovation was not confined to Britain. America might not have been under the rule of Queen Victoria, but its inhabitants were still capable of coming up with devices that typified the era's reputation for ingenuity.

In 1899, the Chicago and Alton Railway Company built what they called 'the handsomest train in the world'. Needing a picture of the train that would do justice to its magnificence, the company called in Chicago photographer George R. Lawrence. He suggested that, for the handsomest train in the world, he would shoot the largest picture in the world. And for that, he would build the world's biggest camera.

It was a bold statement, but Lawrence ran a studio whose slogan was 'The Hitherto Impossible in Photography is our Specialty'. If anyone could do it, he could.

George R. Lawrence, the Chicago photographer who inspired the world's largest camera.

The camera's creator

In 1888, at the age of twenty, Lawrence already had a reputation as an inventor. He had built a telegraph system, made a gun in his own forge and was fascinated by building devices to simplify household tasks including an elementary form of washing machine.

In 1891, he opened a photographic studio in Chicago. Despite knowing little about chemistry, he began experimenting with flash photography, which in those days was performed by use of explosive powder ignited at the moment of exposure. Undeterred by explosions and accidents that burned off his hair, eyebrows and moustache, he eventually developed a formula for a flash powder that gave more light with less smoke than before, earning himself the nickname of Flashlight Lawrence.

Lawrence was also a pioneer of aerial photography, first from high towers, then from a cage attached to a balloon – an enterprise that nearly killed him when the cage became detached 200 feet up, plummeting him to the ground, his fall broken only by telephone wires. Following the accident, from which he escaped unharmed, Lawrence moved on to shoot aerial pictures using a camera attached to a kite, or sometimes a string of kites.

Lawrence, undoubtedly, was a special type of photographer, which was no doubt why the Chicago and Alton Railway Company called him in to photograph their new train. It was like no other. All windows were of the same size, shape and style. Every carriage was the same size, length and height. The coal tender was made to the same height as the carriages behind it. They called it the *Alton Limited*, and ran it from Chicago to St Louis.

These were the days when cameras were mostly made of wood, and very few used rolls of film. Instead, pictures were taken on glass plates, coated with the photographic emulsion, loaded into the camera and shot one at a time. With his interest in panoramic photography, Lawrence's initial thought was to take a series of pictures, to be joined during the printing process. The railroad company, however, having built the perfect train, didn't want an imperfect picture with joins in it.

So Lawrence proposed building an enormous camera, then using it to shoot a picture on a single glass plate measuring 8 x 4½ feet – larger than any other photographic plate ever exposed. The camera was to be called the Mammoth. He got the go-ahead.

Building and moving the camera

To manufacture the camera, Lawrence turned to J.A. Anderson in Chicago. The company was already making wooden studio cameras and only three years before had made the Anderson Tailboard Camera, which could almost have been a miniature model for the Mammoth. The tailboard design featured a lens panel at the front, linked by bellows to a ground glass screen at the back, which moved on a track to focus the lens. The picture was first focused on the screen, and then the plate holder was placed into position for the exposure. What Lawrence needed was a much larger version of a similar design.

Engineers at work assembling the Mammoth.

The Mammoth was made at the Anderson works, using cherry wood. It had two focusing screens, made from celluloid, which moved back and forth like sliding doors. The bed of the camera was 20 feet long and supported on four 2 x 6-inch cherry beams. The bellows, which moved on steel track wheels, were in four sections – two square and two cone-shaped – and made of rubber, with each fold stiffened by a quarter-inch slat of wood. Two more canvas linings inside made it lightproof. More than 500 feet of wood and 40 gallons of cement were used to build the bellows alone.

The plate holder was equipped with a roller blind made from 80 square feet of ash wood, lined with three thicknesses of lightproof material to cover the plate's light-sensitive surface prior to the exposure. It slid up and down on ball bearing rollers, spaced every 2 inches, which reduced friction to such an extent that a 14-year-old boy could operate it.

Wide-angle and telephoto lenses were built to a German Carl Zeiss design and were the largest photographic lenses ever made. The wide-angle had a focal length of 5½ feet and the telephoto focal length was 10

Loading the camera onto a horse-drawn van for transport to the railway station.

feet. To get an idea of the scale of such a thing, consider a more conventional plate camera of the time, for which a wide-angle lens was more likely to have a focal length of about 2-3 inches.

The photographic plates were made by the Cramer Company in St Louis, who also manufactured the sensitised paper on which the plate's negative image would be contact printed to make a positive. Coating the plates with the light-sensitive emulsion involved laying them flat on a huge marble slab, beneath which sheets of ice rapidly cooled the emulsion as it was applied. The plates cost $1,800 for twelve.

The Mammoth took eight months to design and build, at a reputed cost of $5,000. With its plate holder fitted, it weighed 1,400 lbs and took 15 men to operate it.

The camera was made in sections, and, like smaller and more conventional cameras, its size could be reduced by folding the bellows, from the full 20 feet extension down to 3 feet. Even so, it took at least half a dozen men to manhandle it into the back of a horse-drawn wagon that took it to the Chicago and Alton Railway Station. There, it was

The camera being transported to the shooting location by a railway flat car.

transferred to a flat car, pulled by one of the company's locomotives. Fifteen men, many in suits, starched collars and bowler hats, climbed aboard and the train made for Brighton Park, about 6 miles from Chicago, where the camera was assembled a quarter of a mile from the tracks where the train was waiting to be photographed.

Taking the picture
It took at least six men to manipulate the bellows and focus the lens on the screen at the back, and four men to get the huge photographic plate in its holder onto the back of the camera. But before the exposure was made, the plate had to be checked to ensure that no dust or dirt had accumulated during its journey to the site.

For this, the camera's front panel, which supported the lens, was opened and a man stepped inside. The door was closed and a red filter placed over the lens. Since the photographic plate was insensitive to red light, the roller blind could then be safely rolled up so that the man could see, by the red light from the lens, to remove any dust or dirt from the plate. The roller blind was then rolled down, the lens filter removed and the front panel opened for the man to exit.

Assembling the camera ready for the big picture shoot.

Taking the picture with the Mammoth. George R. Lawrence is the man standing beside the lens.

Dismantling the camera after the picture had been taken.

A lens cap was placed over the lens and the roller blind once more rolled up to expose the sensitive surface of the plate to the capped lens. On a bright and sunny day in the spring of 1900, Lawrence made the exposure by removing the lens cap for two and half minutes and then putting it back into place. The roller blind was rolled down and the plate holder removed. Transported to a darkroom, the plate was then developed and three contact prints made from it. The process took 10 gallons of chemicals.

The prints were sent to the Paris Exposition of 1900, where they were at first thought to be fakes, and the French Consul General was sent from New York to Chicago to inspect the camera. When he saw it, he became convinced of the pictures' authenticity.

As a result, the French Exposition officials awarded George R. Lawrence, user of the world's biggest camera and photographer of the world's largest picture, the Grand Prize of the World for Photographic Excellence.

The picture the camera was built to take.

CHAPTER 8

The First Channel Tunnel

The idea for a tunnel under the English Channel between England and France had been proposed long before the first successful attempt opened in May 1994.

As far back as 1802, French mining engineer Albert Mathieu-Favier put forward a proposal for a two-bore tunnel between the two countries, the top bore used for horse-drawn stage coaches and the lower one used for drainage and groundwater flow. The next year, Englishman Henry Mottray proposed a tunnel made from prefabricated iron sections. By the 1830s, with the arrival of steam railways, the idea for a tunnel began to gather momentum and French mining engineer Aimé Thomé de Gamond began a thirty-year project in which he worked on seven different designs.

By 1855, Queen Victoria began taking an interest. During a state visit to France that year, she and French Emperor Napoleon III approved one of Gamond's designs and it was presented to the Exposition Universelle of Paris in 1867. Just as with the other previous proposals, nothing came of it.

It wasn't until 1880 that the first real attempt at digging a tunnel between the two countries really took shape. The man behind it was Sir Edward Watkin, a Member of Parliament who, as was the case with so many Victorian entrepreneurs, had strong connections with the emerging railway infrastructure. In fact, he was chairman of nine British railway companies, which in turn led him to various railway projects, of which a Channel tunnel was just one.

Work begins
Work began in 1875, when Acts of Parliament in both England and France authorised the Channel Tunnel Company Ltd to begin trials. In 1877, two shafts were sunk, one in France at Sangatte, and a second one in England

THE FIRST CHANNEL TUNNEL

Sir Edward Watkin, the man behind the first Channel Tunnel.

at St Margaret's Bay, near Dover. Unfortunately, flooding forced it to be abandoned.

In 1880, under the direction of Sir Edward Watkin, a new shaft was sunk between Folkestone and Dover, with a horizontal tunnel bored along the cliff above the high-water mark. The workforce consisted of Welsh miners, who tunnelled for 800 feet before a second shaft was sunk at Shakespeare Cliff, near Folkestone. From here, it headed out under the sea with the intention of meeting the French pilot tunnel midway across the English Channel.

Entrance to the tunnel as it was in 1984 – subsequently sealed up and secured by a steel door.

Inspecting the boring machine.

Tunnel workers being lowered down the shaft.

At the time, there was some confusion about the machinery used. It was originally envisaged that the work would be carried out by a compressed-air boring machine that was invented and built by Colonel Frederick Beaumont, who had been involved with the Channel Tunnel Company since its start. But in 1880, Captain Thomas English, from Kent, patented

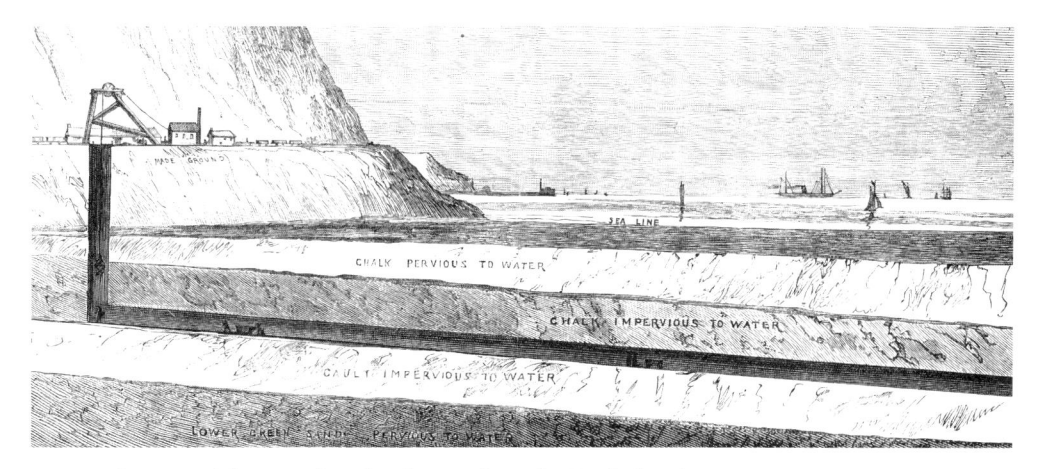

Section of the tunnel under the sea from the English side.

The bottom of the shaft, where the boring machine was lowered and assembled.

a better and more efficient machine. Beaumont, however, was credited with having his machine used, when a report appeared in *The Engineer* magazine; the mistake was never corrected, hence the confusion.

Captain English's machine was 33 feet long. Powered by compressed air, it used a rotating cutting head that could cut five-sixteenths of an inch with every rotation. Rotating at two revolutions per minute, it was capable

The boring machine used to dig the tunnel.

Looking towards the tunnel from the bottom of the shaft.

THE FIRST CHANNEL TUNNEL

of boring through the ground at the rate of half a mile per month. As it moved, men shovelled debris into buckets, which automatically shot their contents into a skip. Once full, the skips were hauled manually along rails to the mouth of the tunnel, where they were emptied.

The machine bored a circular tunnel, 7 feet in diameter, which would be excavated by a second machine behind it, enlarging the bore to a diameter of 14 feet.

The plan was for the tunnel to be mostly circular, with a flat bottom for railway lines, on which locomotives ran by compressed air. Charged with the air before they started out, there would be enough power to pull 250 tons weight along 22 miles of tunnel. In the absence of ventilation shafts, which would be part of the design of a conventional tunnel under land, rather than the sea, the air discharged by the engine would provide enough to sustain the passengers. In case of emergencies, reservoirs of air would be set at intervals along the tunnel length for recharging the locomotives if necessary.

It was thought that the chalk seabed through which the tunnelling took place was suitably stable, but to prevent falls of chalk, the tunnel was lined with 2 feet of concrete, made from shingle taken from nearby Dungeness, the largest shingle beach in Europe. With a 1:8 gradient, the tunnel was planned to reach 150 feet below the seabed before levelling off.

Looking east along the tunnel.

THE INGENIOUS VICTORIANS

With the French boring their part of the tunnel from the other side of the Channel, the expectation was that the two tunnels would meet by 1886, 11 miles from the Kent coast. Some time before that, however, things began to go wrong.

Problems arise
First indications that the tunnel might be in difficulty came when funds from the British Government to finish the 11-mile stretch failed to materialise. Undeterred, Watkin set up a new company, called the Submarine Continental Railway Company, in 1882, and prepared a Bill to put before Parliament.

It fell on deaf ears. By then, various military sources had begun suggesting that a tunnel would make it easy for a Continental army to reach Britain and take the country by surprise. Watkin countered that should that be the case, the tunnel could be easily flooded, ventilation restricted and lift cables cut to prevent invaders from exiting the tunnel. The fact that there was political instability in France at the time didn't help Watkin's case.

A section of the flooded remains of the 1880 tunnel.

THE FIRST CHANNEL TUNNEL

Watkin battled on to preserve his tunnel. He invited top businessmen and prominent citizens, including the Lord Mayor of London, to inspect the work so far, and held a luncheon party in the tunnel. Later, the Prince of Wales and the Archbishop of Canterbury were among the notable visitors to be invited.

Despite all Watkin's efforts, the government was determined to put a stop to his work. One way was by invoking an Act of 1881 that forbade tunnelling below the low-water mark without permission, demanding work should stop immediately. Watkin talked his way out of that one and it was agreed that tunnelling could continue subject to an inspection by the Board of Trade. The result of the inspection was a recommendation that tunnelling should cease. Watkin ignored instructions and continued. When another visit by Board of Trade officials revealed that more tunnelling had taken place, court proceedings were taken out against Watkin's company.

Work ceases

In the years that followed, many Bills were put before Parliament, not just from Watkin's own companies, but from new rivals equally determined to build a tunnel. All were thrown out, and in 1898, Watkin was permanently restrained, by the Chancery Division of the High Court, from further boring beneath the seabed.

The shafts to the tunnel were infilled and lay forgotten until as late as 1974, when work begun on another tunnel at the same site resulted in

A miss-spelt inscription on the wall near the start of the tunnel reads 'This tunnel was begubnugn in 1880 William Sharp'.

the rediscovery of the first tunnel. That project was also abandoned. The last evidence of the original Victorian tunnel was discovered when it was bisected by the boring machinery of the present Channel Tunnel, when work began in 1988.

Today, an entry to the tunnel in Shakespeare Cliff, near Folkestone, can still be seen, although the wooden door to it has been replaced by a secure steel barricade. Beyond it, some sections of the Victorian tunnel still remain and have been explored, but they are not accessible to the general public.

Edward Watkin was clearly a visionary man who, had it not been for the political obstructions put in his way, would most probably have succeeded in being the first to tunnel from England to France. Maybe one of the nicest testimonials to his work is an inscription on the tunnel wall, found by much later explorers, about 250 yards from the start. Left by one of the Welsh miners, including a badly corrected spelling mistake, it reads: 'This tunnel was begubnugn in 1880 William Sharp'.

CHAPTER 9

England's Answer to the Eiffel Tower

When the Eiffel Tower was opened in Paris in 1889, it was the tallest man-made structure in the world, erected as the impressive entrance to L'Exposition Universelle, France's World's Fair which, that year, was celebrating the centennial of the French Revolution. Named after Gustave Eiffel, whose engineering company built the tower, it soon enjoyed great success with famous visitors who included Britain's Prince of Wales, the French stage and film actress Sarah Bernhardt, American Wild West showman Buffalo Bill Cody and American inventor Thomas Edison.

It also inspired four English copies: one smaller version, which succeeded spectacularly; two others that enjoyed short runs of success; and a fourth version planned to be even taller than the French structure – and which failed dismally. But first, a few facts about the Eiffel Tower to see what English engineers were up against if they wished to build their own versions.

On completion, the Eiffel Tower stood 1,063 feet high and weighed 10,100 tons. It consisted of four cantilevered legs, anchored firmly to the ground and meeting at the pinnacle. Construction was from more than 18,000 pieces of wrought iron, held together by about 2.5 million rivets. Lifts operated between the ground and three platforms, which contained international restaurants, a theatre, laboratories and a small apartment where Gustave Eiffel entertained guests. In the year it was opened, blue, white and red lights, rotating from the top, illuminated the sky, while searchlights below swept the city and its surrounds up to more than 6 miles away.

Edward Watkin's vision
One Englishman who believed that he could build a structure to not only imitate the Eiffel Tower, but in fact to better it, was Sir Edward Watkin, the man whose interest in the railway structure of Victorian Britain had,

Six of some of the sixty-eight designs entered for the Great Tower for London design competition.

a few years before, led him to an attempted construction of the first Channel tunnel.

Another railway inspired scheme led him to purchase land in Wembley, where he intended to build an amusement park with the hope of enticing Londoners out into the countryside, by way of his Metropolitan Railway line. The park was planned to include waterfalls, ornamental gardens, boating lakes, cricket and football pitches; and at its

ENGLAND'S ANSWER TO THE EIFFEL TOWER

centre would stand Watkin's answer to the Eiffel Tower. Many names were suggested for it, among them The Great Tower For London, Watkin's Tower, Wembley Tower, Wembley Park Tower, The Metropolitan Tower, Watkin's Stump and, inevitably, Watkin's Folly.

Since it was inspired by the Eiffel Tower, Watkin initially invited Gustave Eiffel to design the English version, but the Frenchman declined on the grounds that his countrymen might think the worse of him. So, in 1890, an architectural design competition was held. In all, sixty-eight entries were submitted from architects across Britain, mainland Europe and as far afield as America.

Some bore a resemblance to the Eiffel Tower, though many deviated completely with design concepts that ranged from something resembling a tall, extremely thin wigwam to a series of ever-diminishing towers that stood one on top of the other. One design looked a lot like a gigantic screw. There was a proposal for a copy of the Leaning Tower of Pisa, and another that suggested a tower with a spiral railway line climbing up and around its exterior. Perhaps the most bizarre idea was for a tower that included a scale model of the Great Pyramid of Giza, in which vegetarians might live above the ground, growing their food in hanging baskets.

Competition entry number 59, an unusual tripod design put forward by Professor Robert H. Smith and Mr William Henman.

A failed design

On 20 June 1890, *The Engineer* magazine went into print with details of one of the more attractive entries, maybe in anticipation of it being the winner. It was design number 59 in the competition, submitted by civil engineer Professor Robert H. Smith and architect William Henman, who illustrated and described their design in the magazine.

Unlike the Eiffel Tower, which stood on four legs, their tower was conceived as a tripod design, standing on three steel tubes, 30 feet in diameter and 350 feet apart at the base, tapering to 6 feet in diameter and only 25 feet apart at the top.

Standing 1,400 feet high and weighing 11,500 tons, it would be more than 300 feet taller than its Paris counterpart, and weigh 1,400 tons more. Its estimated cost was £229,000.

The bases of the steel tubes were to be tied together underground. At 156 feet above the ground, they would be tied again and a gallery erected. Above that, at 282 feet from ground level, the first great platform was to be built with a floor space of 54,385 square feet. Moving up to 558 feet from ground level, a second platform was proposed, with a floor area of 12,580 feet. At the top of the tower, a third platform would boast a floor area of 4,075 feet. It was reckoned that the total amount of floor space on the three platforms could accommodate 8,520 people in comfort. If all areas were crowded, the capacity might rise to 20,000 people.

Inside each of the three steel tubes, two-deck lifts were planned to ascend from the ground to the first platform. Although they would run on curved rails, the engineering was such that their floors remained horizontal at all times. They would be driven by hydraulic rams, or water pumps, to lift and lower the lifts on wire ropes, and would be capable of transporting 2,000 people per hour.

Travel between the first platform and the second was to be by means of a single, circular-shaped, glass-walled cage that rose in the space between the steel tubes. Its capacity was for up to 1,000 passengers per hour.

The journey from the second platform to the top of the tower was planned by means of another two-deck lift operating in the space between the tubes, with a capacity of 800 people per hour.

For those who preferred to walk, there would be staircases inside the tubes, one for those going up and two for those coming down. The staircases would be lit by light from glazed windows, and end at the base of the mast at the very top of the tower.

Right from the start, it was intended that this particular tower should look graceful while retaining an impression of height, grandeur and strength. To keep to that precept, it was decided to employ a simplicity of design without any discernible breaks in the steel tubes, and with all the necessary bracing and ribbing on the insides. The windows would be inset into the tubes so that they wouldn't interrupt the smoothness of the design. The result, in the picture shown in *The Engineer* magazine, was a cleanliness of line that made the tower look elegant and retained its intended gracefulness. The added advantage was that the smoothness of the exterior meant less wind resistance.

Maybe Smith and Henman, in writing about and illustrating their concept in such detail in *The Engineer*, thought they might take a leaf out of the book of Sir Joseph Paxton, who, in publishing his plans for the Crystal Palace in *The Illustrated London News* before they had been officially accepted, managed to sway public opinion in his favour.

The winning design

The plan might have worked for Paxton, but not so for Smith and Henman. Their vision for the tower was one of the more elegant designs, but it didn't win. Instead, the 500 guinea prize went to London architects

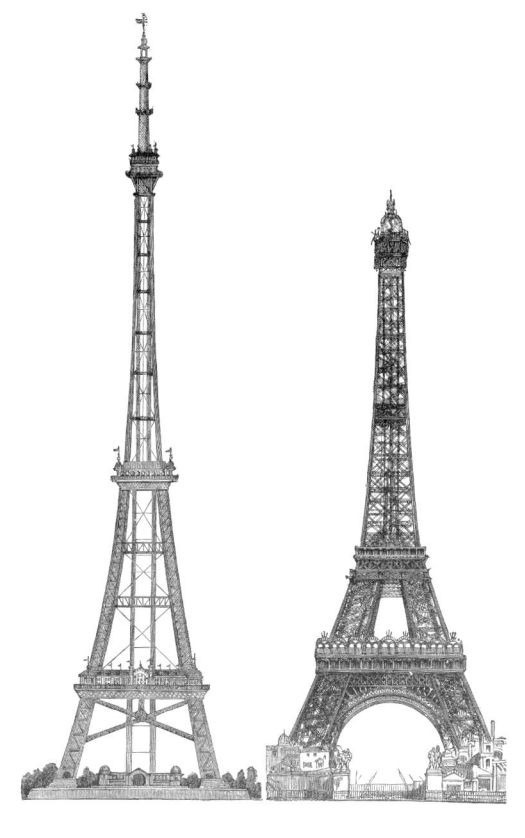

An artist's impression of how the finished tower was planned to look, shown in scale with the Eiffel Tower (right).

THE INGENIOUS VICTORIANS

Stewart, MacLaren and Dunn, with a design made from mild steel, planned to stand at 1,200 feet and bear a much closer resemblance to the Eiffel Tower.

To attain the stability required for such a project, the winning tower was an octagonal shape, a design that was known to offer the least resistance to bending from all directions. At its base, it was 300 feet in diameter. Four lifts were planned up to the first stage, along with two staircases in the tower's legs.

The principal stage was 200 feet above ground and contained a large octagonal-shaped hall, 20,000 square feet in size and 60 feet high. A surrounding balcony contained shops, restaurants, sideshows, and a hotel with ninety bedrooms. The floors were made of concrete and steel.

The second stage, 500 feet above the ground, was reached by a further three lifts, and housed another hall of 10,000 square feet, as well as another selection of shops and restaurants. Higher still, a third platform housed a post office, a telephone call office and other small buildings. An observatory was planned for the very top, with a searchlight to shine out over the surrounding landscape.

All the lifts ascended vertically, rather than having to travel partly up the tower's legs, and were designed to carry 60,000 people per day.

Wembley Park, with the tower shown as it was envisaged it might look when completed.

Girders about to be hoisted into position.

The whole tower would be lit by electricity and would weigh 14,659 tons. The estimated cost was £352,000.

All the materials used were brought to the site by railway, whose lines ran to each of the four legs. A steam winch was used to raise the four legs. Thereafter it was planned for the rest of the tower to be built, using four electric cranes, each weighing about 20 tons, and running on girders bolted to the sides and above the tower's horizontal struts. The cranes would unload direct from the railway trucks, and as each section was completed, it was estimated that the cranes would be able to lift themselves to the next level in less than half an hour. Being electric, they were powered by a central dynamo, which had only enough power to operate one crane at a time. By the time the tower reached 500 feet, it was decided only one of the cranes would be needed to continue the work, sliding between the lift guides.

It was expected that the tower would be finished and open to the public by 1895. Things started out auspiciously, but went rapidly

The first stage of the tower under construction in April 1894.

downhill. To manage the project, Watkin formed the International Tower Construction Company, and employed Benjamin Baker, a civil engineer who had been involved with the design of Scotland's Forth Bridge and the Aswan Dam in Egypt, to oversee construction. The first major blow came when Watkin appealed for a public subscription to help fund building. The public, however, were far from forthcoming, and the company fell back on using its own funds.

To reduce costs, Watkin commissioned a redesign that would use four legs, more like the Eiffel Tower, rather than the eight originally envisaged. Foundations were laid in 1892 and work began the next year. At the same time, preparations began on the cricket pitch, boating lake and other amenities planned for the parkland surrounding the tower. The park opened to the public in May 1894, and proved to be a great attraction for Londoners. Its centrepiece, however, was a very long way from completion. It was September 1895 before even the first stage of the tower was built, standing at 154 feet like a large platform in the air supported on four legs.

End of the dream
At this point, ill health forced Watkin to retire. Not long after that it was discovered that the tower's foundations were insecure. Financial problems followed, and in 1899, the construction company went into voluntary liquidation.

ENGLAND'S ANSWER TO THE EIFFEL TOWER

Watkin died in 1901, with work on the tower never having got beyond that first stage. The site was declared unsafe for the public, and work began on the demolition, using dynamite to blow it up, in 1904.

The park continued to attract visitors. It was later home to the British Empire Exhibition of 1924, whose sports area eventually became the Wembley Stadium of today. When the stadium was rebuilt in 2000, the pitch was lowered, and the original foundations of the tower were rediscovered.

The tower in Wembley was not the only one inspired by the Eiffel Tower. There were three others, all in the north of England.

New Brighton Tower

In 1896, the New Brighton Tower was built at Wallasey, in Cheshire, 567 feet tall with its base surrounded by a four-storey red brick building. It had a ballroom and theatre, housed exhibitions and had grounds that contained a boating lake, funfair, gardens and sports ground. Advertising itself as the highest structure in the UK, visitors were charged one shilling (5p) for a single entrance, or could sign up for a summer-long season ticket for ten shillings and sixpence (52½p).

Four lifts took visitors from the ground to the top in ninety seconds, but one night in 1909, a woman and child were stranded at the top all night, when staff checks failed to notice they weren't aboard the final lift of the day, which descended without them.

The tower also featured a ballroom with sprung floor, offering six hours of dancing, six days a week. It was visited by English composer Edward Elgar, who conducted his famous *Enigma Variations* musical piece at the Tower in 1898.

The New Brighton Tower.

The New Brighton Tower was closed in 1914 with the outbreak of the First World War, after which it grew rusty and was dismantled when the war ended.

THE INGENIOUS VICTORIANS

The East Promenade of Morecambe, with the newly built Tower in the backround, in Victorian times.

Morecambe Tower

In 1898, the Morecambe Tower was built on the coast of Lancashire to stand at 232 feet high. It had a pavilion that seated 5,000 spectators. With the outbreak of the First World War, however, it was demolished and the steel used for munitions.

Blackpool Tower

It was another tower built a year after the Eiffel Tower, six years before the New Brighton Tower and eight years before the Morecambe Tower, that eventually survived the Victorian era and went on to achieve worldwide fame. It was built at Blackpool, in Lancashire.

Despite being inspired by the Eiffel Tower, the Blackpool Tower stood at only 518 feet high, a little less than half the height of its French counterpart. It also differed by having its base embedded in a bricks and mortar building.

The tower's origins began in 1890, with the acquisition by The Blackpool Tower Company, of Dr Cocker's Aquarium, Aviary and Menagerie, which had been on the town's promenade since 1873. Here,

ENGLAND'S ANSWER TO THE EIFFEL TOWER

the company built its English replica of the already famous French landmark. Dr Cocker's establishment remained open during the building and was then incorporated into the tower.

Blackpool Tower was designed by local Lancashire architects James Maxwell and Charles Tuke, both of whom sadly died before its opening in May 1894, nearly three years after the foundation stone had been laid in September 1891. The total cost was £290,000. Keeping with local materials, the tower used 5 million Accrington bricks, a special type of extra hard engineering brick made in Lancashire; 2,500 tonnes of iron; and 93 tonnes of cast steel.

The building at the base housed a circus, with an admission price of sixpence (2½p). Admission to the tower itself cost another sixpence, with a further sixpence to ride the hydraulic lifts to the top. Right from its earliest times, the tower's ballroom was one of its great attractions. Known as the Tower Pavilion when it opened in August 1894, it was smaller than the subsequent Tower Ballroom, built three years later. The ballroom floor was made from 30,602 blocks of mahogany and walnut, and covered 14,400 square feet.

Blackpool Tower in Victorian times.

THE INGENIOUS VICTORIANS

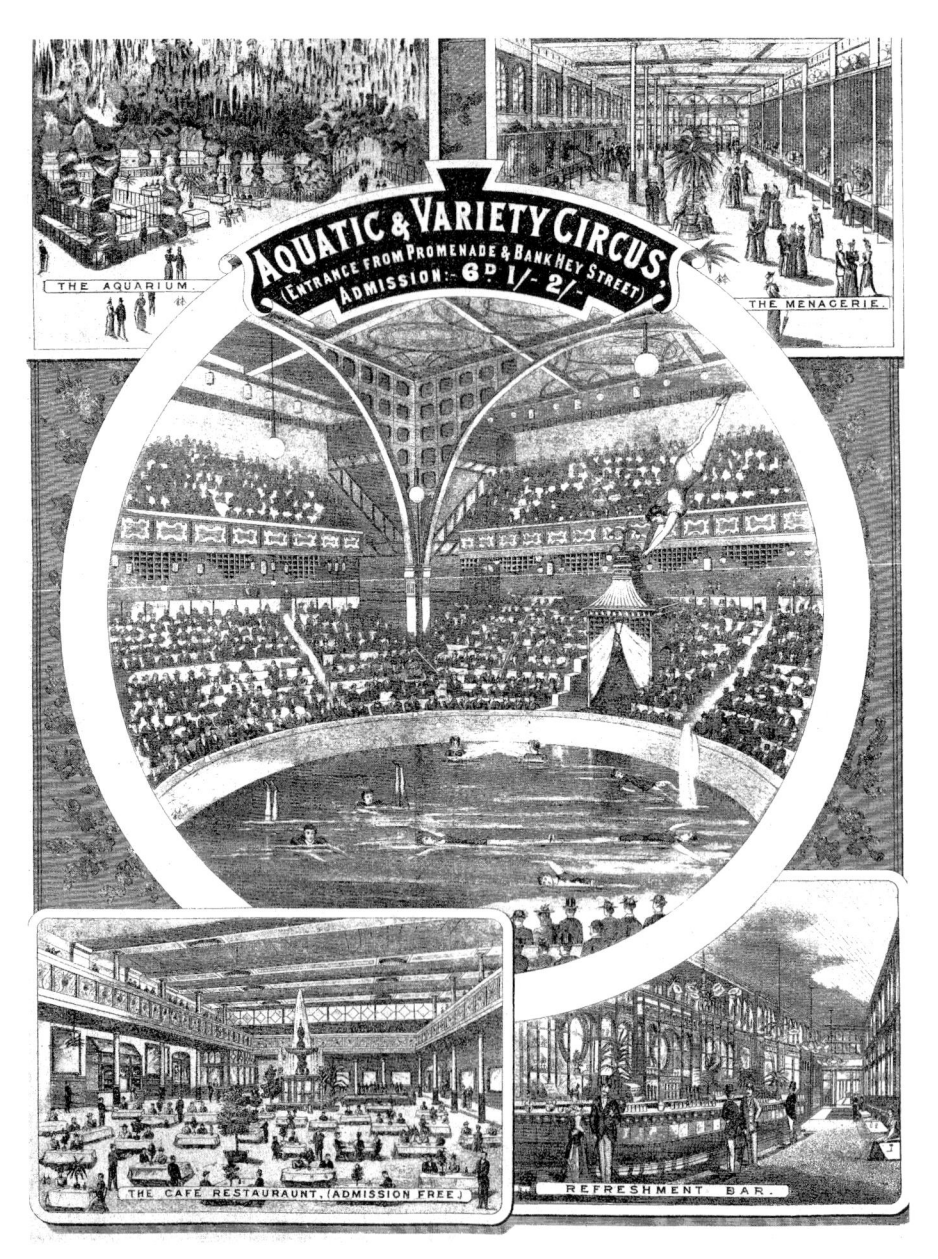

Attractions of Blackpool Tower, as shown in a Victorian programme for the venue.

ENGLAND'S ANSWER TO THE EIFFEL TOWER

The building of Blackpool Tower in 1893.

In Victorian times, dancing in the ballroom wasn't as free and easy as it would later become. There were strict rules stating that disorderly conduct would mean immediate expulsion, and also instructions that gentlemen may not dance unless with a lady.

Of the four English towers envisaged or built in answer to France's Eiffel Tower, only Blackpool Tower has survived, and is still open for business today. Two of the other three were moderately successful for a while. But, inevitably, it's the one that failed that, for historians, is the most interesting. Had it succeeded, it might have been the biggest success of them all.

CHAPTER 10

Under, Over and Across the Thames

London was a thriving port in Victorian times, and the river Thames, as it flowed through the City and on to the North Sea, was both a blessing and a burden. The blessing was that it was deep and easily navigable for the tall-masted vessels needed to bring goods right into the centre of the capital from Britain's coastal regions and overseas. The burden was that it formed a natural barrier between North and South London, creating problems for those who needed easy ways to cross the river, especially cargo carriers from the expanding docks that lined both banks.

Bridges across the Thames were the obvious answer, and had been for a great many years. The first stone-built London Bridge, as opposed to previous rickety wooden structures prone to catching fire and burning down, dated back to 1176, and until 1750, was London's only Thames bridge. By the Victorian era, the Pool of London, which stretched from London Bridge east to Limehouse, was where the supply ships docked. But because their masts were too tall to sail under any normal bridge, it was impossible to span the river in a conventional way anywhere east of London Bridge.

The Victorians, however, found ingenious ways to make the crossing. They went under the river, with a tunnel that was the first of its kind; they went over the river, with the help of a revolutionary style of bridge; and they went across it, with ferry boats of a particularly unusual design.

The Thames Tunnel
The Thames Tunnel, from Rotherhithe on the south bank of the river to Wapping, on the north side, was the brainchild of Marc Brunel, father of the better-known engineer, Isambard Kingdom Brunel. It was the world's first tunnel to be dug under a navigable river, through soil that consisted of soft sand, clay, gravel and quicksand. Begun in February 1825, twelve

years before Victoria became queen, work continued through the early years of her reign.

The mechanics of the building began in an unconventional way, starting with the sinking of a huge shaft, 50 feet in diameter, 150 feet back from the south bank of the river. The conventional way of excavating such a shaft would have been to simply start digging into the ground, but Brunel had a better idea. He built the shaft above ground in the form of a circular 3-foot wide brick wall, 42 feet high, on top of a sharpened steel rim, looking more like a huge tower. It weighed about 1,000 tons, and once built, the whole thing was allowed to simply sink into the ground under its own weight. Inside, men at the base worked to excavate the earth that built up as the tower sunk, whilst a steam engine at the top pumped out water that filled the bottom from the wet ground. It sank at the rate of a few inches per day.

When the tower, which was rapidly becoming a shaft, had sunk to within a mere 2 feet of its intended stopping point, it became jammed. Brunel solved that problem by turning off the steam pumps, thus allowing water to filter in and soften the ground, and then by adding 50,000 extra bricks to the top to give it more weight. In this way, it completed its journey until the top of the tower was at ground level. The process took three months.

With the shaft in position, it was now time to begin boring horizontally under the river. The machinery used for the task was inspired by a type of worm called *Teredo navalis*, which Brunel had encountered while working at Chatham Dockyard. The worm bored its way into the timber of wooden ships using shell-like protrusions either side of its head, ingesting the wood and then excreting it from its body, using it to reinforce the tunnel it was creating.

In Brunel's mechanical version of the worm-inspired technology, a huge shield was erected, containing thirty-six man-sized chambers, twelve across and three high. Each of these was open to the rear for a workman to enter, but closed at the front, with moveable boards that butted up against the earth where tunnelling was to take place.

The miners were burrowing through a thick sludge that made it impossible to dig too much at one time for fear of the excavation collapsing around them. To combat this, they dug the earth out in layers. First, the top row of boards was removed and digging commenced to a depth of just a few inches. The short excavations this created were shored

The three divisions of the shield used by the miners.

How the shield was used to dig into the earth ahead of the bricklayers building the tunnel behind it.

up and the boards fixed at the ends, up against the newly exposed walls of earth. Then the operation was repeated with the next row of boards below, followed by the lowest row. When each of the thirty-six excavations had been made in this way, the cage was moved forward on screw jacks and the whole process begun again. Behind them, bricklayers worked to build the walls of the tunnel. In this way, they eventually bored 1,200 feet, literally inches at a time.

In the years that followed, the tunnelling continued slowly, with Isambard joining the team in 1826 at the age of twenty. The next year, there was the first flood, when the workmen dug too close to the river bed. Marc Brunel borrowed a diving bell from the East India Docks Company and descended from a boat to repair the damage.

Once the water and silt were cleared from the tunnel, Isambard Brunel organised a banquet in the tunnel to restore confidence among the

A diving bell is used to examine the tunnel damage from above.

The stairway that led down to the twin tunnels, illustrated to show it running beneath the Thames.

workforce and the public. Fifty influential gusts were invited to dine at a huge table, while the Band of the Grenadier Guards deafened everyone with music whose volume was amplified by the acoustics of the tunnel.

With confidence restored, work began again, but in 1828, there was a major flood that filled the tunnel with water. Six miners died and Isambard came close to being killed, escaping only when the tidal wave of water swept him back along the tunnel to the entrance shaft, where it rose to the surface, taking the young Brunel with it.

The tunnel was pumped out, but it was seven years before work began again. Financial problems and a lack of confidence in the project led to the tunnel being bricked up in 1828, seemingly never to be completed.

But Marc Brunel was not one to give up on his dream. By 1834, he had managed to raise the cash to reopen the tunnel, and in 1835, work began again. Further floods, fires, gas leaks and explosions, not to mention major financial problems ensued, but work carried on until, in August 1839, the Wapping shore was reached, where a shaft similar to that on the Rotherhithe side was sunk.

On 25 March 1843, eighteen years after work began, the Thames Tunnel was opened to the public and proclaimed the Eighth Wonder of the World. It had originally been envisaged as a tunnel for horse-drawn vehicles, but lack of finances prevented the building of spiral ramps in

A souvenir medal of the time, showing Marc Brunel on one side, and the tunnel on the other.

The boiler house at Rotherhithe that once contained steam-powered pumps to extract water from the tunnel, today houses the Brunel Museum.

the shafts that would have been needed for the purpose, and so it was used only by pedestrians.

In 1865, the tunnel was sold to a railway company, and steam trains began running under the river. The steam trains have gone, but the tunnel remains today in use by the London Overground railway.

Tower Bridge
In 1894, Victorian ingenuity found a new way to build a bridge over the Thames, east of London Bridge. The difference between this and any previous bridge was that the road, which conventionally carried foot passengers and vehicles, was unconventionally split in the middle, so that its two sides, or bascules, could be raised or lowered like twin drawbridges to allow tall-masted ships to pass beneath. It was positioned close to the Tower of London and was called Tower Bridge.

The conception of the bridge began in 1877, with the launch of a public competition to find a solution to the ongoing problem of crossing the river whilst still allowing ships to reach the Pool of London.

One proposal was based on a design first put forward by mechanical engineer and naval architect Sir Samuel Bentham back in 1801, for a way of improving the Port of London with a new look at London Bridge. In his design, the bridge widened in the middle, with a road on each side of a central pool. A ship approaching one side of the bridge was allowed through into the pool by the operation of horizontally moving swing bridges in one road. These were then closed and a second pair opened in the road on the opposite side of the bridge to allow the ship to move the rest of the way and out the other side. In this way, road traffic would not be interrupted by the passage of the ship.

It would seem that Bentham failed to patent this idea, because eighty-four years later, in 1885, Frederio Barnett of Queen Victoria Street, London, was granted a patent for a 'bridge for facilitating the passage of shipping', which followed identical principles.

One problem with this type of bridge was that it required long piers to support the swinging portions of roadway, something that could be overcome by making the moving parts of the bridge lift vertically, rather than swinging horizontally. This thinking led to another design concept, proposed by City architect Sir Horace Jones, reporting on the various projects of civil engineer Sir Joseph Bazalgette. His specification showed two towers with suspension bridges leading up to them and with twin

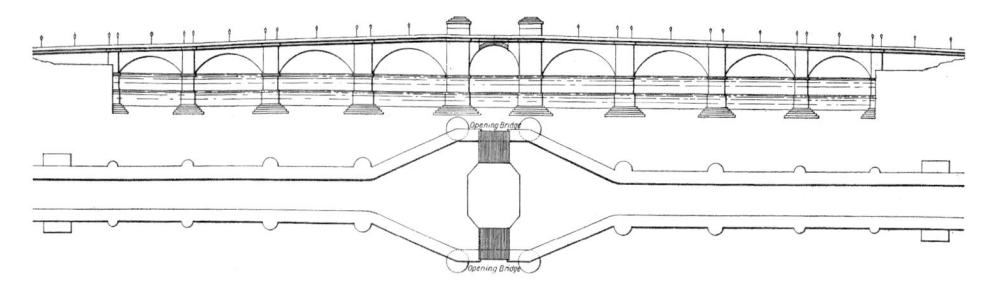

From the viewpoint of an approaching ship and from above: Samuel Bentham's design of 1801.

bascules between them. Closed they would span the river 29 feet above high water – the same as London Bridge – with a 300-foot span that split and opened to a height of 130 feet above the water.

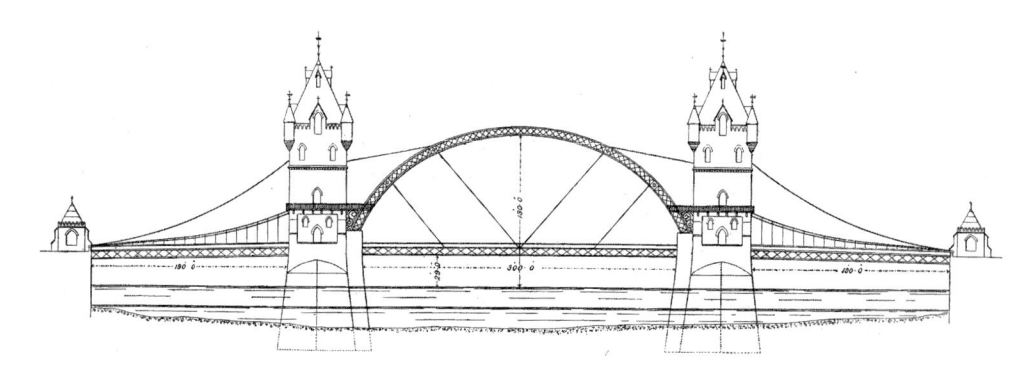

Sir Horace Jones's proposed design.

After seven years of deliberation and the consideration of further designs, the committee came to a decision. Approval went to a design by civil engineer Sir John Wolfe Barry.

In Barry's winning design, the total length of the bridge was 800 feet, and for 270 feet on each side it took the form of a suspension bridge. Two towers, each 213 feet high, stood either side of the 200-foot section in the middle, and it was this section that was designed to divide into two bascules, each of which weighed more than 1,000 tons and lifted to allow ships to pass through. Suspended between the tops of the towers was a pedestrian walkway.

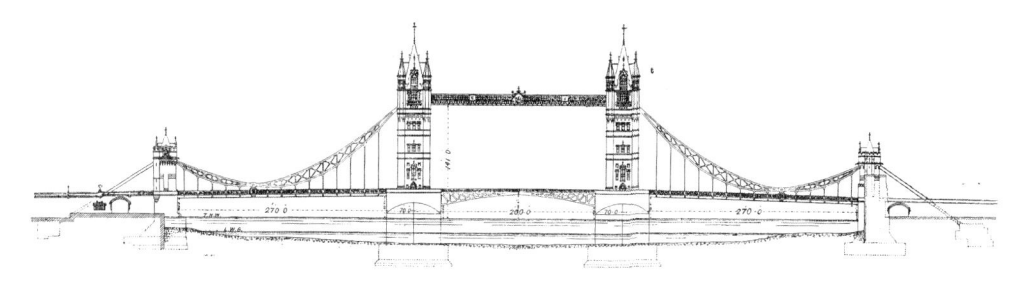

The winning design from Sir John Wolfe Barry.

The lower parts of the great piers in place in the river.

The bascules were counterbalanced to make the raising and lowering easier. The mechanism was powered by hydraulics, in which two steam engines pumped water under pressure into pressurised storage reservoirs known as hydraulic accumulators. These, in turn, fed the driving engines that raised and lowered the bascules.

Construction began in 1886, and took eight years to complete, involving 432 workers. More than 7,000 tons of concrete were sunk into the riverbed to support the bridge, and more than 11,000 tons of steel went

Work begins on one of the towers.

Riveting the top of a large girder that would support the bridge.

UNDER, OVER AND ACROSS THE THAMES

Gangs of workmen hoist the heavy ironwork of the piers into position.

The bridge begins to take shape.

The opening of Tower Bridge in 1894.

The arrival of the royal party.

into the towers and walkways. At the time, it was the largest and most sophisticated bascule bridge in the world.

Tower Bridge was officially opened in 1894 by the Prince of Wales, later to become King Edward VII on the death of Queen Victoria. The royal party arrived and travelled across the bridge in a horse-drawn carriage, between ranks of naval and military guards of honour and to the sound of the National Anthem, played by the Guards' band. The opening ceremony was performed on a dais, at the front of which stood an ornate plinth, surrounded by flowers at its base and with a lever on top. The Prince announced in loud, clear tones that, in the name of the Queen, the bridge was now open for land traffic. He then turned the lever, which was linked to the bridge's hydraulic machinery, and, said *The Graphic* newspaper of 7 July 1894, 'the bascules, watched with the greatest interest by all the Royal visitors, began to rise with slow dignity.'

A guardsman on one of the bridge's balconies then used a flag to signal the firing of a ten-gun salute on the shore, and the royal party departed from the bridge by a steam boat called *The Palm*, which took them to Westminster.

The Prince of Wales pressing the lever that raised the bascules.

The ten-gun salute that followed the official opening.

Souvenir pamphlet from the opening ceremony of Tower Bridge.

To control the passage of river traffic through the bridge, semaphore signals were used during the day and a system of red and green lights at night to indicate when the bridge was open or closed. In fog, a gong was sounded as well. Vessels passing through by day were required to display a large black ball of at least 2 feet in diameter, high up on their masts. At night, twin red lights were required to be displayed in the same position. The ship's steam whistle was also employed in fog, when passing through the bridge.

Open and closed: Tower Bridge towards the end of the Victorian era.

UNDER, OVER AND ACROSS THE THAMES

The reaction from the public to Tower Bridge was mixed. Architect and author Henry Heathcote Statham, writing in 1916, was quoted as saying: 'It represents the vice of tawdriness and pretentiousness, and of falsification of the actual facts of the structure.' The artist Frank Brangwyn also commented: 'A more absurd structure than the Tower Bridge was never thrown across a strategic river.'

Needless to say, neither of its detractors very much harmed the reputation of Tower Bridge. It is still in use by an estimated 40,000 people every day, and river traffic still takes precedence over pedestrians and vehicular traffic. The bascules are raised more than 1,000 times a year, although the original steam-powered mechanism has been replaced by an electro-hydraulic system.

Woolwich Ferry

There were numerous small ferries operating across the river Thames in Victorian times, but the best of them was at Woolwich, historically part of Kent, but absorbed into the County of London in 1889. It was a town split by the river Thames.

There had been ferries at Woolwich since the 1300s, when the town was a fishing village. From then on, various companies ran boats between the two banks of the Thames until, in 1884, the Metropolitan Board of Works agreed to provide a free ferry for passengers, animals, goods and horse-drawn carts. It opened to the public, to great acclaim, in March 1889.

The Woolwich Ferry boats were of a peculiar design that looked more like Mississippi paddle steamers than the kind of boats that might have been expected to be steaming across the Thames in East London. They featured two decks – foot passengers on the lower one, vehicles on the upper – with two tall chimneys fore and aft, and a huge paddle wheel on each side. The captain's bridge was built high across the width of the ferry, with a corridor beneath, through which passengers could walk and watch the engines at work.

The engines were coke-fired and stoked by hand to avoid excess smoke, each one using about 8 tons of coke a day. They were known as diagonal surface condensing engines, and each ferry had two pairs, with each pair connected to one paddle. In this way, one paddle could turn forward, while the other paddled in reverse as an aid to turning the boats, which also had rudders fore and aft. The result was that the ferries were

The first generation of ferries at Woolwich: the Gordon *and the* Duncan.

extremely manoeuvrable, enabling them to steam across the river in either direction before turning sharply to meet the jetties side-on. They were lit by incandescent electric lighting.

The ferries all had similar specifications: 172 feet long, 62 feet wide, a tonnage of 625 tons and a speed of 8 knots. Each could carry up to 1,000 foot passengers and twenty vehicles.

UNDER, OVER AND ACROSS THE THAMES

The outer edges of the iron bridges that linked the land to the ferries, and by which passengers and vehicles embarked and disembarked, were supported by pontoons, which allowed the structures to rise and fall with the tide. The brows, which gave vehicular access from the bridges to the ferries, were raised and lowered by hydraulics.

Two ferries operated at first, both purpose-built in 1888 by a company called R.H. Green. One was called the *Gordon*, named after General Charles Gordon, who was born in Woolwich in 1833 and died in 1885 defending Khartoum against Sudanese rebels. The second was the *Duncan*, named after Colonel Francis Duncan, who wrote a history of the Royal Artillery, which was based at Woolwich until 2007.

Woolwich ferries crossing the Thames at the time of the official opening.

On the opening day, flags and bunting hung from every building in the south bank town, and the streets were lined with soldiers from Kent regiments, most of whom were based at the Royal Arsenal in Woolwich. Mounted police led a procession of dignitaries through the streets and onto the *Gordon*, waiting on the south bank of the river. It took the ferry three and a half minutes to make the crossing to the north bank, where the passengers were met by another procession, more bands and the fire engine of the Beckton Gas Works, specially decorated for the occasion. Making the journey back again to the south side of the river, an official

The third ferry, introduced in 1893, was the Hutton.

ceremony was performed by Lord Rosebury, Chairman of the London County Council, who declared the ferry open and free for ever. The day ended with a banquet for 200 people at the Freemasons' Hall.

In 1893, a new boat of similar design joined the fleet, built by William Simons and Company. Called the *Hutton*, this one was named after Charles Hutton, a professor of mathematics at the Woolwich Academy.

The Woolwich Ferry service lasted long past the Victorian age, with very little change to the basic principles. In 1922, two new boats replaced the original ferries, this time built by Samuel White and Company, but once again following the paddle steamer design of the earlier ferries. The first of these was the *Gordon II*, again named after General Gordon. The second was called the *Squires*, in honour of William James Squires, twice

Mayor of Woolwich and one-time chairman of the Woolwich Building Society.

To complete a quartet of ferries, Samuel White and Company built two more to join the fleet in 1930. The first was the *John Benn*, named after Sir John Benn, an ancestor of Member of Parliament Tony Benn and a one-time member of the London County Council. The second was the *Will Crooks*, named after William Crooks, also a member of the London County Council, Mayor of Poplar and Woolwich's first labour Member of Parliament.

In all, the ferries ran almost continuously for seventy-four years, only the thickest of fogs halting the service, even through the years of the Second World War. It has been estimated that during that time, they clocked up about 400,000 miles and carried 18 million passengers. They were sold for scrap to Jacques Bakker and Zonen in Belgium in 1963, and replaced by three diesel vessels.

CHAPTER 11

Shocking Cures for Every Ailment

The Victorians did not discover electricity, but they certainly put it to good use in everything from lighting to electric cars. For many, however, electricity was very much a mystery whose powers were misunderstood and often misinterpreted.

Not least among the more dubious applications for electricity was its use in the medical world, where even qualified medics were of the opinion that a good electric shock was the cure for a great many ailments. Electricity also proved to be a godsend to the quack doctor fraternity who, in the days long before advertisements were required to be completely factual, sold devices that claimed to cure everything from backache to baldness, and a whole lot more in between. At the end of the nineteenth century, advertisements abounded for electric helmets, corsets, belts, hairbrushes, socks and even complete suits.

Electro-therapy
Magneto-electric therapy machines were among the more popular devices sold to cure medical ills. To quote the label inside one such device: 'This machine has been designed expressly for the use of the medical profession and for invalids who are unable to take exercise, suffering from rheumatism and various other complaints.' According to the many machines on the market, other complaints that might be cured by electro-therapy included cancer, tuberculosis, diabetes, gangrene, heart disease, tetanus, spinal deformities, toothache, neuralgia, debility, sleeplessness, sciatica, lumbago, torpid liver, organic weakness and other kindred ailments.

Some machines were in the form of heavyweight appliances that would have been difficult to move around, but the most popular types were portable and usually contained in a compact wooden box. Turning a crank on the outside of the box wound a magneto inside that used

SHOCKING CURES FOR EVERY AILMENT

The Improved Magneto-Electric Machine of Daniel Davis, 1848.

magnets to produce bursts of alternating current. Attached by wires to the magneto were two metal handles, usually made of brass, which were placed in the patient's hands. Alternatively, one might be held, whilst the other was applied to areas of the patient's body that were appropriate to the ailment being treated. Whichever way it was used, it was important for the patient to be in contact with both handles so that an electronic circuit was completed through the body. Here are the directions printed on the inside lid of the box of one such machine:

Connect two metallic cords or wires with the sockets in the ends of the box and apply the handles connected with the other ends of the metallic cords or wires to any part of the person through which it is desirable to pass the current of electricity. Then turn the crank, regulating the strength of the current by the speed, and by the knob at the end of the box; it being desirable to increase the strength only to that degree most agreeable to the patient. It is less unpleasant to the patient if wet sponges are placed in the ends of the handles, and these applied to the skin, as they prevent any prickling sensation. The sponges should never be put inside of the box while wet as they rust the machinery. In applying it for toothache, tic-doloureux, or neuralgia, the operator takes the handle and places his fingers or sponge over the part affected, while the patient holds the other handle. In applying it to the foot, place one of the handles in the water with the foot, and hold the other in the hand, or apply it to any other part of the person. The bearings and spring must be oiled occasionally.

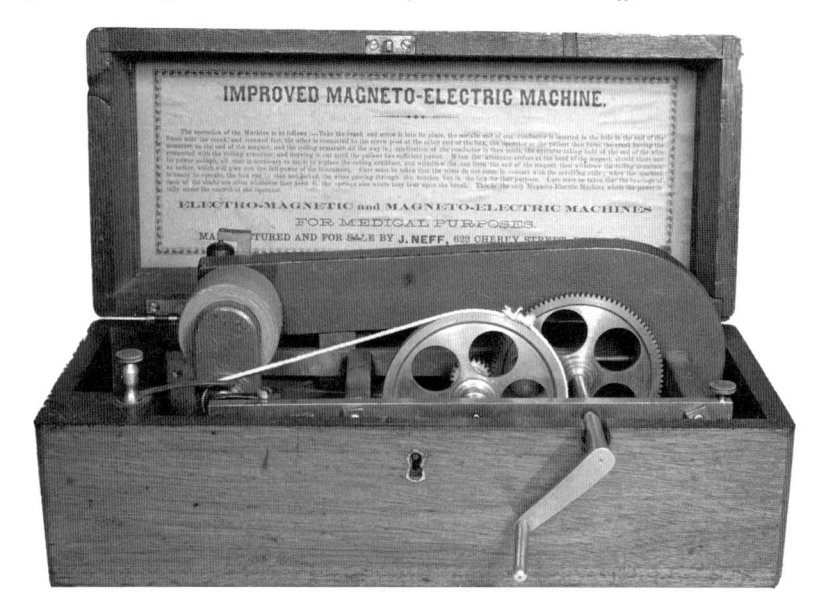

A portable Magneto-Electric Machine from W.C. and J. Neff, 1870.

SHOCKING CURES FOR EVERY AILMENT

In many so-called cures for various ailments, Victorian patients might have complained of feeling no immediate effect from the treatment. That was something that no one treated by magneto-electric therapy could ever complain about!

Electric belts and corsets
In the strictest sense of the word, many so-called Victorian electric garments were magnetic rather than electric. But that didn't stop their makers and retailers from using the magic references to electricity in their advertising.

There was, in fact, a modicum of scientific truth behind some of their claims. Experiments had already shown that when rods of copper and zinc were connected by an electrical conductor and placed respectively into copper sulphate and zinc sulphate solutions, the resulting chemical reactions caused electrons to pass along the conductor from the zinc rod to the copper rod in the form of a very weak electric current.

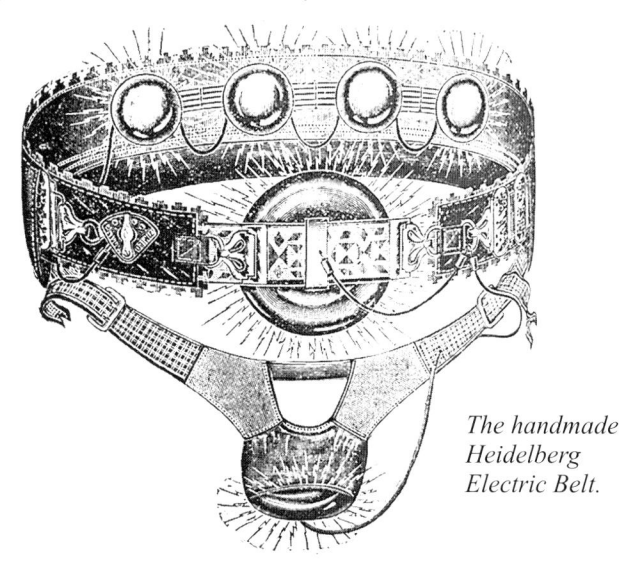

The handmade Heidelberg Electric Belt.

This was something that obviously hadn't escaped the attention of Cornelius Bennett Harness, proprietor of The Medical Battery Company in London's Oxford Street. He was one of many retailers who sold electropathic belts for men, containing zinc and copper plates that were claimed to produce a health-giving current. Many such belts, looking unnervingly like a kind of electronic jockstrap, never actually claimed to be a cure for impotency, but it wasn't difficult to read between the lines of the advertising of the day to see what was implied.

Cornelius Harness was also the driving force behind the electric corset. The patent for the garment, granted to George A. Scott, an Englishman who made his reputation in America, spoke about the use of copper, zinc and highly magnetised steel within the corset, but which might also be applied to any garment or article of underwear, the closer to the skin, the better. The object was to get as much metal as possible in close proximity to the body. The bands of different metals were brought into contact, said the patent, to secure voltaic action. Harness began as a distributor for Scott's corsets, before making his own version, which he sold for five shillings and sixpence (27½p).

Harness's advertising claimed: 'The very thing for ladies, young or old, especially those who suffer from a weak back, chills, rheumatism, hysteria, internal complaints, loss of appetite, nervous dyspepsia, kidney disorders, rheumatic and organic affections, ladies' ailments, biliousness, etc.'

Not only that, but it gave its wearers a better figure: 'By wearing this perfectly designed corset, the most awkward figure becomes graceful and elegant, the internal organs are speedily strengthened, the chest is aided in its healthy development and the entire system is invigorated.'

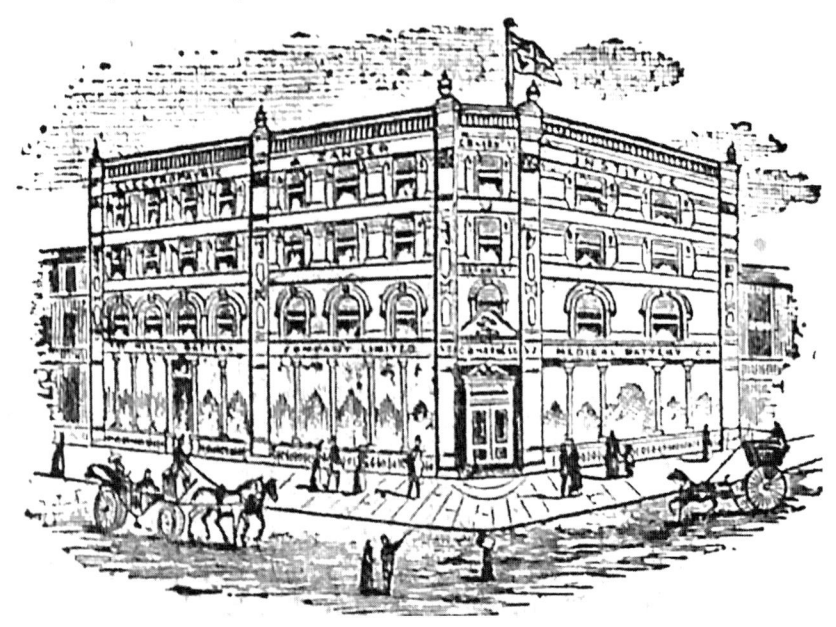

Premises of the Electropathic & Zander Institute, in London's Oxford Street.

The cover of a pamphlet for the Electric Corset, from the Medical Battery Company.

A pamphlet from Harness, full of glowing testimonials from women who claimed to have been cured of ailments ranging from a weak back to constipation, explained that every married man should buy one for his wife, if he valued her health, comfort and appearance.

The electric corset seems to have been the flagship of Mr Harness's empire, but it was by no means the only electric medical device sold by his company. From the same premises in Oxford Street, he operated the Electropathic & Zander Institute, for the relief and cure of disease by electricity, massage and Swedish mechanical exercise.

His field of expertise also took in an electro-dental department for painless dental operations, and tooth extractions for half a crown (12½p). Also available were washable, self-adjusting trusses for every form of rupture and internal prolapse. For ladies troubled by superfluous hairs, the Institute offered removal by the electrolytic action of electricity; for the deaf, there was an electricity based cure that worked without the slightest pain, inconvenience or danger – or so it was claimed.

A cure for baldness
If you don't want to go bald, or want to promote extra hair growth, brush your hair with an electric hairbrush. That would seem to have been the message behind one electronic device, aimed equally at men and women of the Victorian era.

One of the more prolific makers of electric hairbrushes was George Scott, he of the electric corset patent, but referred to in his hairbrush advertising as Dr Scott. 'Will not save an Indian's scalp from his enemies, but it will preserve yours from dandruff, baldness and falling hair,' claimed Scott's advertisements.

As with some other contemporary devices, Scott claimed that his brushes were driven by electricity, when in fact what power they did exert came only from magnetism, by incorporating magnetised iron rods in their handles. The most obvious advantages, he claimed, was that use of his brushes would increase hair growth, whilst also giving relief from headaches. Rather incongruously, he also claimed that brushing the hair with one of his brushes might cure constipation, malarial lameness, rheumatism, diseases of the blood and paralysis.

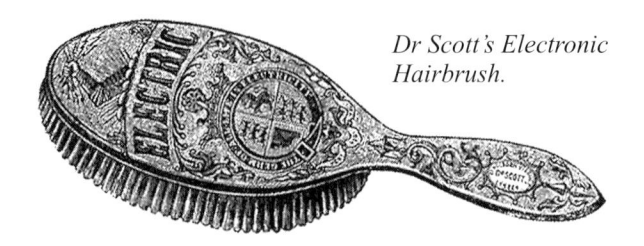

Dr Scott's Electronic Hairbrush.

Scott's prowess as a businessman was also indicated by a warning on the hairbrush boxes that claimed that only one person should use each brush if users wanted their brushes to retain their full curative power. In this way he encouraged families to buy more than one brush.

As well as electric – or magnetic – hairbrushes, Scott also introduced electric insoles, rheumatic rings and plasters. His advertising for the plasters claimed that they combined electro-magnetism with all the best qualities of standard porous and other plasters. In one to three hours, advertisements claimed, use of the plasters would cure colds, coughs and chest pains; nervous, muscular and neuralgic pains; stomach, kidney and liver pains; dyspeptic, malarial and other pains; rheumatism, gout and inflammation. 'They who suffer ache and pain need suffer never more again,' claimed his advertisements.

It was, in fact, possible to buy from Scott practically an entire suit of pain-relieving garments, including shoulder braces, throat protectors, nerve and lung invigorators, body belts, wristlets, sciatic appliances, anklets, leg appliances, office caps, or any other special appliance that could be made to order.

A cure for hysteria
Victorian ladies were prone to suffer from a form of hysteria unknown to affect men. It was an affliction that had been identified as far back as the Ancient Greeks, and possibly beyond. By the middle of the nineteenth century, it was believed that three-quarters of the female population suffered from the affliction, whose symptoms were both physical and emotional. They included headaches, emotional instability, melancholy, aggression, depression, feeling lower abdominal heaviness, muscle pains and other discomforts. These were often considered to be linked to a woman's reproductive system.

Although few would be bold enough to suggest such a thing, and even fewer would admit to it, there's no small possibility that much of the problem lay in sexual frustration, rooted in a belief among Victorian women that the sex act was purely for the sake of procreation, rather than anything approaching enjoyment.

Since long before the Victorian era, there had been many suggested cures for this strange ailment, most of which involved massage or vibration in one form or another. The treatment was initially applied manually by a doctor to the appropriate area of a women's body, resulting in the patient becoming extremely agitated before sometimes passing out. When she calmed down or came round, her hysteria was cured – at least for a while. The eventual introduction of various electronic devices, however, relieved doctors of this chore.

'Vibration is Life,' proclaimed one advertisement of the time, that showed an electric vibrating appliance being innocently applied to the small of a woman's back and her head. There were also devices on the market that could be attached to any normal chair to set it vibrating. Advertisements showed a woman sitting in the converted chair, eyes closed with a look of obvious contentment on her face.

Things took a step forward with the introduction of a new device that was announced as an electromechanical medical instrument to provide a reliable and efficient physical therapy to women believed to be suffering from hysteria. 'Aids that every woman appreciates,' said one advertisement, which alongside a sewing machine, showed a device that was described as a 'portable vibrator with three applicators, very useful and satisfactory for home service'. Now, for the first time, in the privacy of their own homes, women could use these latest wonder devices to cure their own hysteria by the simple expedient of inducing in themselves what the Victorians politely referred to as a paroxysm.

Thanks to electricity, and for the first time in a great many years, female hysteria became less of a problem to the fair sex of the Victorian age.

CHAPTER 12

Photographic Detectives

Detectives were very much a part of Victorian culture. Formal detection had begun in the 1800s with the formation of the Bow Street Runners, but by the time Victoria came to the throne, their reputation had become tarnished through their associations with the criminals they were meant to be pursuing. Following the disbanding of the Bow Street Runners, Scotland Yard's first detective force was formed in London in 1842.

About the same time, in the world of fiction, *Murders in the Rue Morgue* by Edgar Allan Poe was published, featuring the English language's first fictional detective, C. Auguste Dupin; and in 1853, Charles Dickens wrote about a detective investigating the murder of a lawyer in his novel *Bleak House*. In America, Pinkerton's Detective Agency, established in 1850, was gaining in repute, whilst back in England, Sherlock Holmes, surely the most famous of all Victorian fictional detectives, made his first appearance in a story called *A Study In Scarlet*, published in *Beeton's Christmas Annual* at the end of 1887. It was a major factor in encouraging the craze for detective fiction, together with an interest in all things covert.

One of the most popular areas into which the fascination for detectives and their work spilled over was in the world of photography, with the appearance and rise in popularity of detective cameras and disguised cameras. Although similar in some ways, those two types need to be differentiated.

A detective camera was one whose design was basically the same as any conventional camera, but whose size and ease of use meant that it could be easily concealed to take covert pictures. A disguised camera was one that strongly resembled some other, non-photographic device. Objects used to disguise such cameras in the Victorian and later ages included books, cravats, walking sticks, waistcoat buttons, tie pins, parcels, belts, rings, lighters, matchboxes, cigarette packets, pocket watches, wristwatches, pens, hats, handbags, powder compacts, briefcases, binoculars, radios and even guns.

So, a disguised camera might have been termed a detective camera, but one specifically designated as a detective camera wasn't necessarily a disguised camera. The disguise taken by various cameras wasn't necessarily to fool people being photographed. Sometimes it was for practical purposes, as an aid to performing certain specialised forms of photography. At other times, cameras were disguised for no better reason than that the buyers wanted something a little different or unusual.

Gun cameras
The design that proliferated in the early days of disguised cameras was the gun, in its various forms. The model generally accepted as the earliest example of a disguised camera began life in 1862, when an English designer named Thompson took out a French patent for a camera whose shape was influenced by the American Colt revolver. This was in the days before film for cameras was made in rolls, as would be the case some years later. Instead, exposures were made on glass plates, which, in conventional models, were loaded into the camera one at a time and capable of recording just one exposure on each.

Thompson's Revolver Camera, one of the earliest examples of a disguised camera.

Thompson's Revolver Camera, or Revolver Photographique, as it was called when demonstrated to the Société Français de Photographie, featured a circular brass body with a wooden gun-type handle on one side and the lens on the opposite side, made to resemble the barrel of a gun. Between them, a circular brass body contained a disc-shaped photographic plate that was rotated between exposures. The lens slid up and down the body on tracks. At the top of its throw, it lined up with a ground-glass focusing screen, covered by a magnifier in a tube on the opposite side of the body to act as a viewfinder.

PHOTOGRAPHIC DETECTIVES

Then, as a trigger-like catch was pressed, the lens dropped to its lower position, which lined it up with the photographic plate, and the shutter was automatically released to shoot the picture. In this way, the camera shot four successive pictures.

Photography in these days was still in the wet plate era, which meant the glass plates on which the exposures were made had to be prepared shortly before the exposure, used in the camera while still wet, and then developed immediately, often using a portable darkroom. But before long, things got easier for the photographer determined to act covertly, with the introduction of dry plates that could be purchased ready for exposure and developed later. As cameras became easier to use, disguised models became more popular.

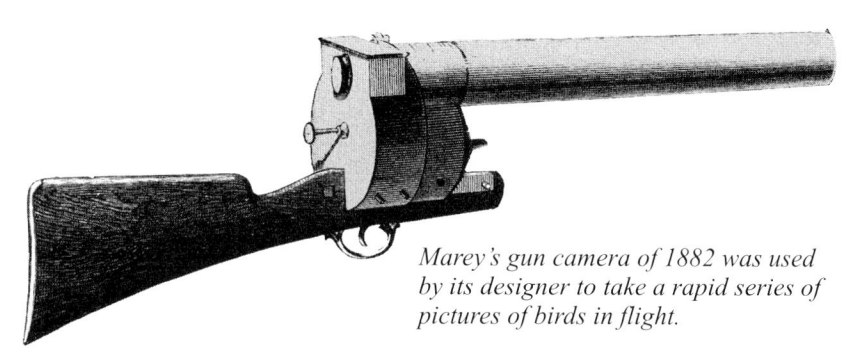

Marey's gun camera of 1882 was used by its designer to take a rapid series of pictures of birds in flight.

Another gun camera, designed by Frenchman Étienne-Jules Marey and made in 1882, was one of those produced for a specific purpose rather than purely to conceal the fact that a photograph was being taken. Marey was studying birds in flight and the camera housed a long-focus lens in the barrel that acted like a telescope to enlarge distant objects. A large circular dry plate was housed in a special magazine and as the gun's trigger was pressed, a mechanism revolved the magazine, to progressively expose twelve pictures in rapid succession.

In 1883, another French camera, called the Photo-Revolver de Poche, took the form of a realistic copy of a European gun of that time, and was made using real revolver parts. It was designed by E. Enjalbert, who went on to design the Postpacket Camera, which looked like a parcel tied up with string, with the lens revealed behind a flap in the end. Another popular design that soon followed produced a series of cameras disguised as binoculars or opera glasses, with one lens for the exposure, the other for the viewfinder.

THE INGENIOUS VICTORIANS

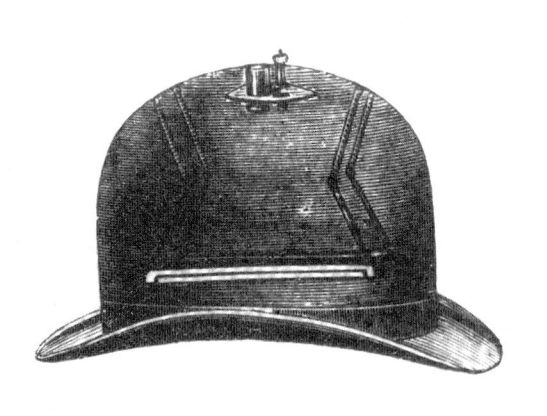

A bowler hat is called into service to house a camera.

Cameras in clothing

In 1886, J. de Neck of Brussels came up with an idea for hiding a camera in a man's hat, with the lens looking out of the front and a viewfinder suspended from the rim. The exposure was made by pulling a string hanging down the side. In a similar vein, K.F. Jekeli and J. Horner came up with a design in which the lens poked out of the top of a hat, which had to be removed from the head and surreptitiously aimed in the direction of the subject in order to make an exposure.

The same year saw the launch of Stirn's Patent Concealed Vest Camera, which took the form of a large metal disc worn beneath a waistcoat (popularly known in America as a vest), with the lens poking through a buttonhole. Six exposures could be taken on a single photographic plate, which was revolved inside the camera using a knob disguised as a waistcoat button. The exposures were made by pulling a string attached to the shutter, and leading from the camera to a convenient trouser pocket.

Stirn's Patent Concealed Vest Camera.

PHOTOGRAPHIC DETECTIVES

One of the more famous designs of disguised cameras came in 1890, with the launch of Bloch's Photo Cravate, sometimes called the Detective Photo-Scarf, in which a camera was hidden inside a large cravat, with the lens disguised as a tie pin. Inside, the photographic plates were attached to a chain that turned to bring each into place for successive exposures.

Cameras in books

The Optimus Book Camera, made by Perken, Son and Rayment, was, as its name suggested, the first of several models made to resemble books, in which the 'covers' were opened to form a triangular-shaped body, with the lens hidden in the spine.

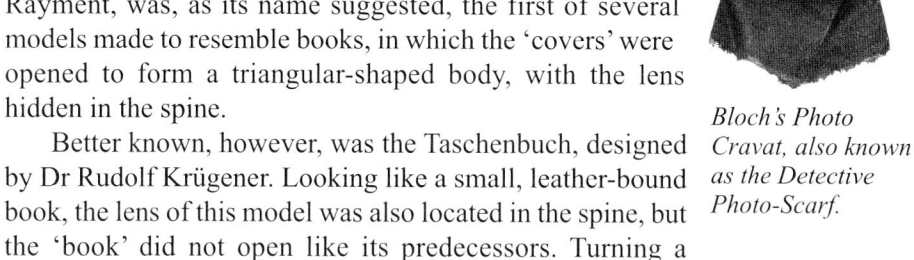

Bloch's Photo Cravat, also known as the Detective Photo-Scarf.

Better known, however, was the Taschenbuch, designed by Dr Rudolf Krügener. Looking like a small, leather-bound book, the lens of this model was also located in the spine, but the 'book' did not open like its predecessors. Turning a handle pushed twenty-four plates individually from a secret compartment into place behind the lens and then on into another compartment after exposure. The camera was sold in several countries, with the book's title, engraved in the leather cover, printed in appropriate languages.

Roll film cameras

When the first roll film snapshot camera was introduced, it didn't take long for the style to be adapted for detective camera use. One such was the Facile, made in 1888 and designed as a simple mahogany box that could be wrapped in paper to resemble a parcel. After the shutter was unobtrusively tripped, a milled knob on the side shifted two grooved boxes inside so that unexposed glass plates in the top box dropped into the lower one, behind the lens, ready for shooting. In the 1890s, photographer Paul Martin used a Facile, wrapped in paper and tucked under his arm, to document life in the streets of London.

The Facile of 1888.

THE INGENIOUS VICTORIANS

Watches, bags and walking sticks

In 1889, English manufacturer Lancaster, known for quality wood and brass cameras, made a significant departure and introduced a new type of disguised camera. The original patent application stated that it could be made in the form of a watch or chronometer case, tobacco box, matchbox, cigar or cigarette case, purse, locket or charm. The most popular design that went into production, however, concentrated on a pocket watch shape, in which the lens sprang out on six telescopic metal tubes, as the 'watch' was opened. The back also opened to allow a small photographic plate to be inserted.

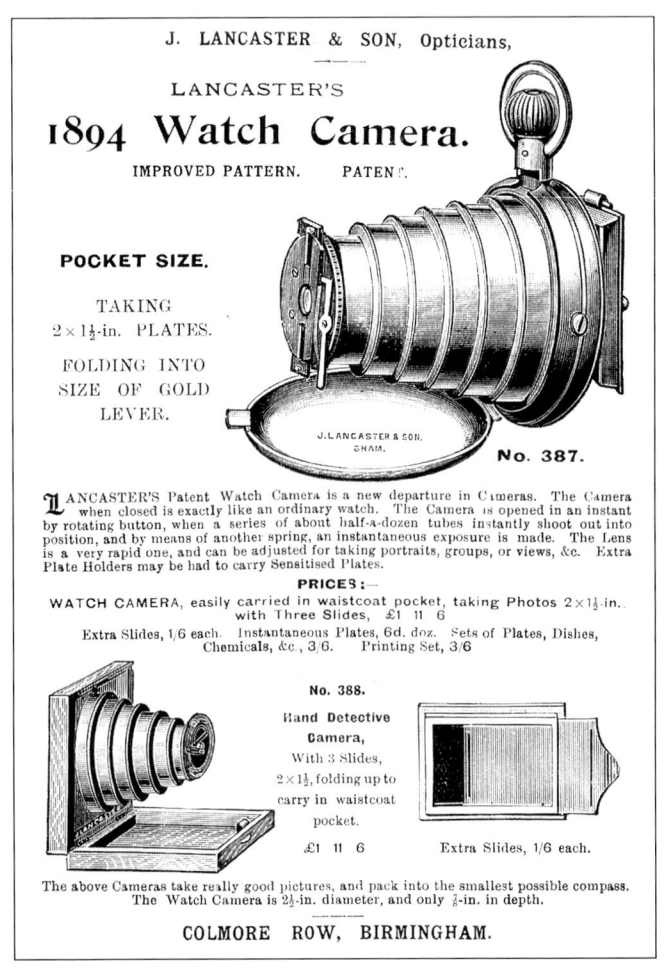

An 1894 advertisement for the Lancaster Watch Camera.

PHOTOGRAPHIC DETECTIVES

Lancaster also introduced the Ladies' Camera in 1894, a fairly conventional design when open, but resembling a lady's handbag when closed, albeit one that was made of wood. Following a similar design, the Photo Sac à Main, made the following year by Charles Alibert in Paris, resembled a metal handbag when closed, but opened to become a plate camera.

Also in 1894, flexible film was at the heart of another watch-shaped camera called the Photoret. Unlike the Lancaster watch camera that preceded it, this model was self-contained in the watch-type casing without the need to unfold it. Film was in the form of a circular disc that took six hexagonal pictures, 13 x 13 millimetres each in size.

The Lancaster Ladies' Camera.

As the Victorian age came to an end, disguised and detective cameras using roll film became more popular, exemplified by the Ben Akiba Walking Stick Camera, invented by Emil Kronke in Dresden. Although

The Ben Akiba Walking Stick Camera.

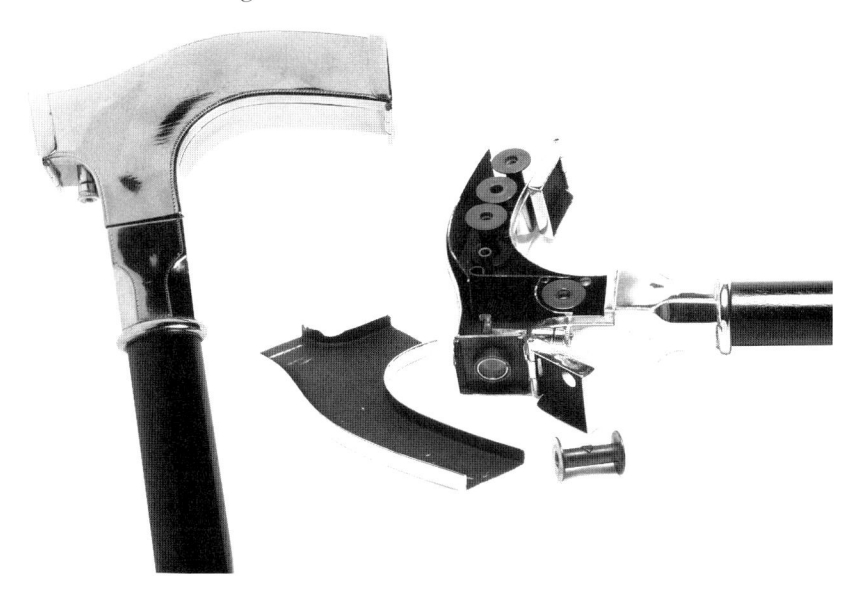

used as a walking stick, the camera itself was only in the handle, with a lens on its front face and a shutter release below it, easily aimed and fired by anyone using the walking stick in the conventional manner.

The solid silver version of the Ticka engraved with Queen Alexandria's monogram.

A few years after the end of Victoria's reign saw the launch of what became the most popular watch-type camera of all time. It was the brainchild of Swedish designer Magnus Niéll, and was similar to the Photoret, but made for roll film rather than its predecessor's film disc. The camera took a small cartridge of film to take twenty-five tiny exposures. The shutter mechanism ran around the inside of the case and the lens was situated in what appeared to be the watch's winding stem, covered by a cap. It was made in America under the name of the Expo Watch Camera, and in England, where it was known as the Ticka. When Victoria died and Edward VII came to the throne, his wife, Queen Alexandria, was presented with special solid silver version engraved with her monogram.

The end of the Victorian era was by no means the end of the craze for disguised cameras, which continued in various forms through the First World War, the Second World War and into the Cold War years, when many were used for genuine espionage purposes. The style remains popular today in the new world of digital cameras.

CHAPTER 13

Frédéric Kastner's Exploding Organ

What do you get when you combine physics, chemistry and music? The answer for one Victorian inventor was a musical instrument called the pyrophone. The clue to how this amazing contraption worked was in the name: pyro, coming from the Greek word for fire, as in pyromaniac, one who cannot resist starting fires.

Georges Eugène Frédéric Kastner was born at Strasbourg in France in 1852 and died in 1882 at the age of thirty. As well as being a physicist and chemist, he was also a musician, being the son of the French composer and musicologist Jean-Georges Kastner. From an early age the young Kastner showed a strong interest in all things mechanical. He was especially interested in the mechanics of railway engines, leading to a fascination for steam, gas and electricity, three motive forces of the Victorian age. Alongside his growing interest in science, he was also tutored in music by his father. Kastner's interests and studies came together when he filed a patent that combined his knowledge of physics, chemistry and music to create the pyrophone, also known as the gas organ, or sometimes the fire organ.

Frédéric Kastner, inventor of the pyrophone.

In any conventional organ – as well as other types of wind instrument – air is blown across a cylindrical tube or pipe to set a column of air vibrating, which, in turn, creates a musical note. In such cases, the air is introduced to the tube or pipe from an external source. The difference with the pyrophone was that the force that started the air vibrating for notes to be created was internal, in the form of a small explosive burst of flame within a tube or pipe.

Musical flames

This highly unusual form of organ was based on a known aspect of physics that recognised the fact that when a flame crosses a glass tube under a certain pressure, it produces a sound akin to a musical note. The theory of combining light with sound in this way was already established before Kastner came onto the scene, with proposals for chandeliers and candelabras that incorporated what were known as singing flames.

In applying this theory to the development of a musical instrument, Kastner, in a patent filed on Christmas Eve 1874, stated:

> It is well known that if a flame of hydrogen gas be introduced within a glass or other tube, and if it be so placed as to be capable of vibrating, there is formed around this flame – that is to say, upon the whole of its enveloping surface – an atmosphere of hydrogen gas, which, in uniting with the oxygen in the air of the tube, burns in small portions, each composed of two parts of hydrogen to one of oxygen, the combustion of this mixture of gases producing a series of slight explosions or detonations.
>
> If such a gaseous mixture, exploding in small portions at a time, be introduced at a point about one-third of the length of the tube from the bottom, and if the number of these detonations be equal to the number of vibrations necessary to produce a sound in the tube, all the acoustic conditions requisite to produce a musical tone are fulfilled.
>
> This sound may be caused to cease either, first, by increasing or reducing the height of the flame, and consequently increasing or diminishing its enveloping surface, so as to make the number of detonations no longer correspond with the number of

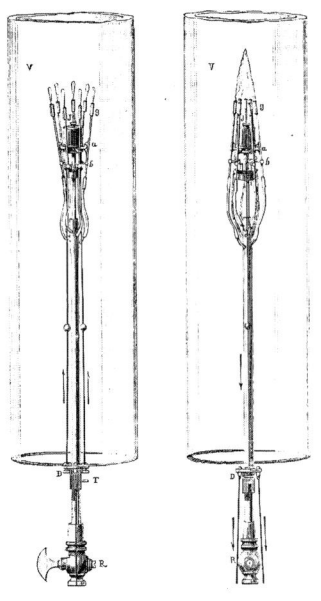

The mechanism that allowed two flames to unite or diverge to produce a musical note.

vibrations necessary to produce a musical sound in the tube, or, secondly, by placing the flame at such a height in the tube as to prevent the vibration of the enveloping film.

Kastner added to this theory with the discovery that two flames introduced into a tube, one third of the way along its length, remained silent when they burned together, but if they were made to separate, the resulting vibrations in the air within the tube produced a musical note.

Taking this thinking a step further, and in order to turn this sound-making device into a musical instrument, Kastner proposed a series of tubes of varying lengths, with holes set at different distances from the top. When a hole was uncovered, the flames burnt together and were silent. When the hole was covered, the flames separated and produced their

The pyrophone, as illustrated in Popular Science Monthly, *August 1875.*

sound. The sounds emitted also changed in tone according to the distance between the flames and the holes. Thus, by controlling the covering and uncovering of the holes by means of a series of keys on an organ-like keyboard, a musical scale could be produced.

The sound of the pyrophone
All of this is a fairly simplified explanation of how the pyrophone worked. The notes and the harmonics produced were also reliant on the size of the tubes and the intensity of the flames. It was suggested that tubes could also be made with a series of holes to be covered and uncovered as needed, in much the same way as holes are covered and uncovered in a wind instrument such as a clarinet or flute. Describing the sound produced by the pyrophone, in 1875, a correspondent in *Popular Science Monthly* wrote:

> The sound of the Pyrophone may truly be said to resemble the sound of a human voice, and the sound of the Aeolian harp [a device in which the wind blowing through strings in a box produced an ethereal sound]; at the same time sweet, powerful, full of taste, and brilliant; with much roundness, accuracy, and fullness; like a human and impassioned whisper, as an echo of the inward vibrations of the soul, something mysterious and indefinable; besides, in general, possessing a character of melancholy, which seems characteristic of all natural harmonies.

Fuel sources to produce the flames in the pyrophone varied. In describing his device, Kastner spoke about the use of hydrogen, but pyrophones were also made to run on propane and even petrol.

Surprisingly perhaps, the reaction to Kastner's instrument was mostly enthusiastic. The eminent nineteenth-century German composer and conductor Wendelin Wessheimer was an enthusiastic proponent of the pyrophone, whilst many scientific journals of the day sung its praises. *L'Année Scientifique* and *Le Journal Officiel de l'Exposition de Vienne* declared that the instrument was 'one of the most original instruments that science has given to instrumental music' and that it could 'produce sounds unknown till the present time, imitating the human voice, but with strange and beautiful tones, capable of producing in religious music the most wonderful effects'.

Kastner's own pyrophone.

Rather more humorously, the *British Review* advocated that the pyrophone might be especially valuable in winter as a means of not only producing music, but also for warming small apartments, and it was suggested that perhaps an economical stove may be added to the instrument for culinary purposes.

Other unusual instruments
Whilst the pyrophone must have been one of the most unusual music making instruments of the age, it was not alone in its weirdness. Xylophones, for example, which had been around for many years, began to come to the attention of musical audiences, although they were played in a vertical upright position, unlike the modern equivalents in which the metal plates that produce the notes when struck are laid out horizontally.

The Aeolia harp was a Victorian update on the Aeolian harp, which dates back to the Ancient Greeks. In the original version, wind blowing through between seven and twenty-one strings contained in a box produced the sound. But, recognising that it was easy for the strings to go out of tune, the Victorians came up with the Aeolia, in which eighty metallic reeds were divided into four groups of twenty harmonic chords. These stood in a box that rotated on a stand in a breeze. The chords were played one after the other as the box rotated and from whichever way the wind was blowing. Their sound was directed into another tube that combined the harmonies.

Nineteenth-century musician Honoré Baudre built a geological piano, whose sound relied on the already known fact that when certain flints are struck together, they produced a musical note. With this in mind, Baudre spent about thirty years travelling the world in search of enough flints to make up several musical scales when struck. According to the pitch of each, the flints were suspended on wires from a metal frame above a soundboard. They were struck by other flints held in the hands to produce the tune.

A similar theory was applied to other rocks and stones that produced various pitches as they were struck. Although the theory had been around for a few centuries, it was in the Victorian era that stonemason Joseph Richardson spent thirteen years collecting enough stones in the Lake District to make his version of the instrument, called a lithophone. Along with his sons, Richardson toured the north of England to give concerts, with a tour that culminated in 1848 with an invitation to perform for Queen Victoria at Buckingham Palace.

CHAPTER 14

Trains in Drains

In Victorian London, horse-drawn carts, cabs and omnibuses all contributed to massive congestion on the roads; the coming of the railway did little to alleviate the problem. The first steam railway passenger service began in 1830, and in the years that followed, mainline stations opened at Euston in 1837, King's Cross in 1852, and Paddington in 1853. A Royal Commission of 1846 had declared that no railway station could be built in Central London, so the only means of transport for passengers and goods between the perimeter-located stations and the heart of the City was by already overcrowded roads. Furthermore, it was becoming apparent that some kind of railway link was needed between Paddington, Euston and King's Cross.

Previous plans for trains to run across the capital on one massive viaduct had been shown to be impractical, largely because of cost, the complications of building and the noise for home owners living near or beneath the viaduct's arches. Also abandoned was an idea for a railway hauled along tracks by miles of hemp ropes pulled by steam engines at each end of the track.

One idea that was briefly considered was for an atmospheric railway, in which carriages were pushed through tunnels by compressed air. Which is where Charles Pearson comes into the story.

Born in 1793, Pearson was the son of an upholsterer, to whom he was briefly apprenticed until he decided to study law. He qualified as a solicitor in 1816, and was elected as a councilman

Underground railway pioneer, Charles Pearson.

of the City of London Corporation in 1817. He was a liberal thinker, a campaigner and a reformist, and in 1847 was elected as a Liberal Member of Parliament to represent Lambeth. In 1845, recognising London's congestion problems, he published a pamphlet advocating a revolutionary idea for what became popularly referred to as 'trains in drains'.

It was Pearson who proposed the atmospheric railway. Like other schemes before it, this one didn't happen. The need for an alternative to road transport within London, however, grew and in 1854, a Royal Commission was set up to look into various new proposals for railways to connect Paddington, Euston and King's Cross. By then, Pearson had abandoned his idea for an atmospheric railway and put forward a new proposal for an underground railway system that would use carriages pulled by steam engines. In August of that year, the Commission gave ascent to the construction of the world's first underground railway, and Pearson threw himself into helping to raise the necessary funds.

The first railway
Far from curing the problem of overcrowded roads, the first method used to build an underground railway made congestion a lot worse, due to the fact that it involved closing stretches of road a section at a time along the route of the railway line. The reason was because, unlike some other tunnels that were bored through the ground, London's first underground railway was built using the cut and cover method. This meant digging a trench along the route, shoring up buildings as necessary on each side, then building up the walls of the cutting and covering it over.

It was a method that had been used before to build tunnels in open countryside. For the London Underground, brick arched roofs to the tunnels were built, which reduced the risk of collapse when the road was reinstated above. It meant that the first Underground lines, built in this way, were close to the surface. Earth-moving machinery, previously developed in the construction of Britain's canal network, was used to remove earth from the trenches, but most of the work was carried out by hand. The width of the first tunnel was 28 feet 6 inches, into which two tracks were laid with a 6-foot gap between them.

Work began on the first line, known as the Metropolitan Railway, in March 1860. The first section, from Paddington to King's Cross, followed a route under South Wharf Road, across Edgware Road and into New Road. From there, it took a straight course to Farringdon. Much of this

Cut and cover works to build the first Underground line at King's Cross.

In the trench at King's Cross.

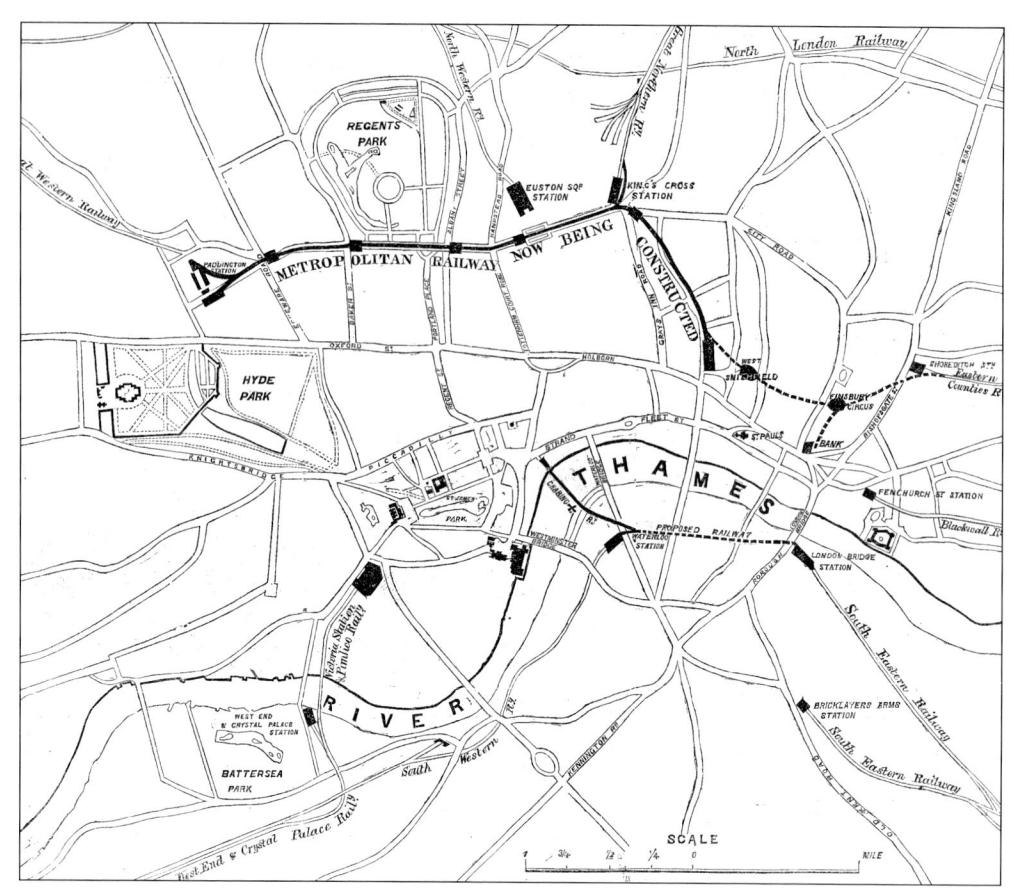

The route of the Metropolitan Railway.

second part of the route was not covered, apart from where it passed under roads that crossed its path.

Where the railway followed the route of an existing road, properties and people were largely unaffected, apart from the inconvenience of road closures during the work. Elsewhere, however, where the route was planned to go under buildings, the cut and cover method meant that any in the path of the line had to be demolished. Estimates of the time suggested about 12,000 Londoners were made homeless as a result.

The stations were situated at Paddington, Edgware Road, Baker Street, Portland Road, Euston Road, King's Cross and Farringdon Street. Stations at each end of the line, at Edgware Road and King's Cross, were open; the others were underground, but spacious and lit by gas.

Construction was not without its problems, including a mainline railway

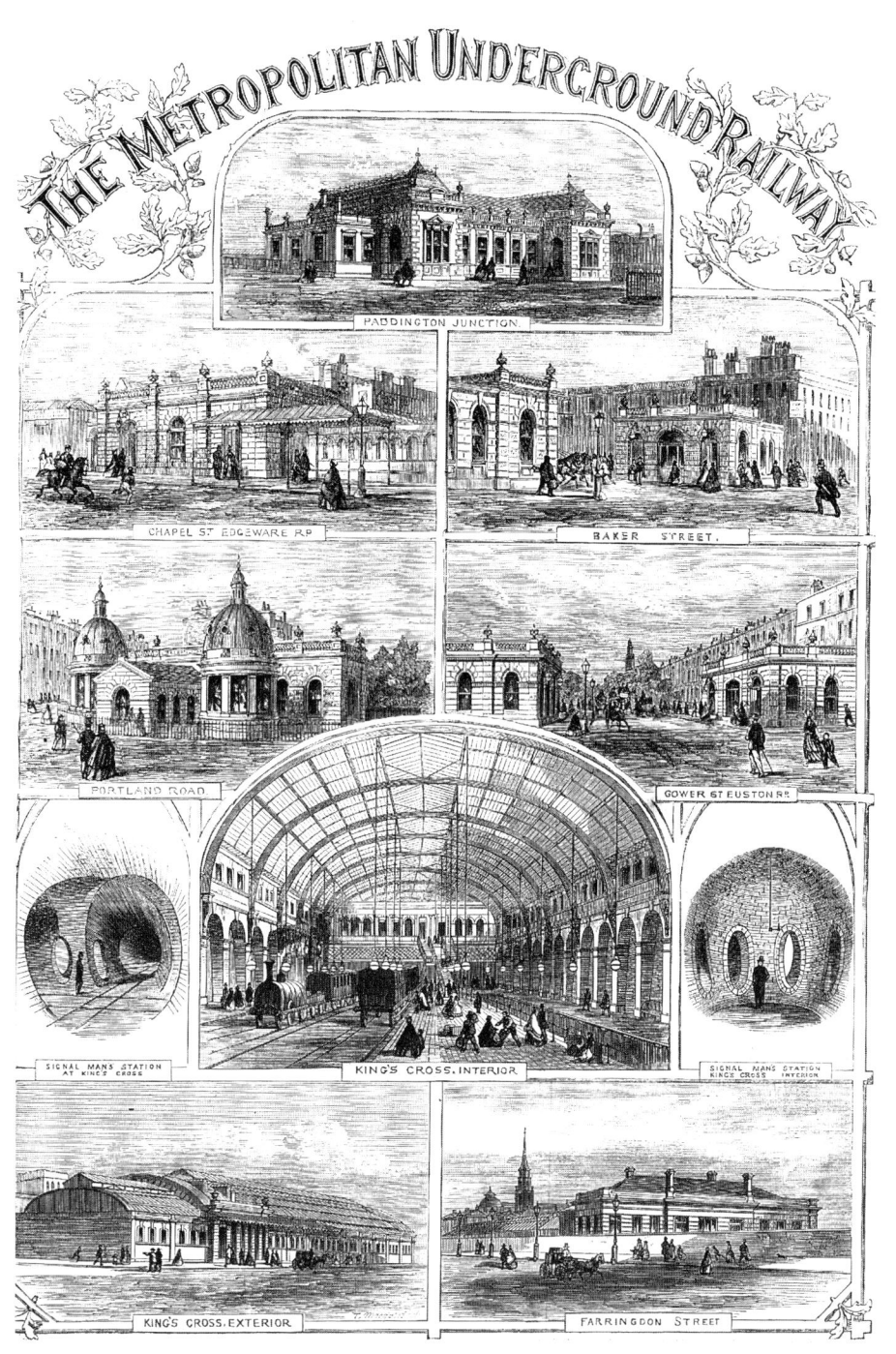

The stations on the first Metropolitan line.

145

engine that overshot the platform at King's Cross and landed in the excavations; a boiler explosion in an engine pulling builders' supplies that killed two people; an excavation collapse that damaged nearby buildings; and a flood caused when a sewer burst. On top of that, there were continual setbacks when the excavation crossed water and gas mains, telegraph wires and even ancient burial pits that dated back to the Plague of London in 1664.

Eventually, a trial run was carried out while the excavation work was still in progress in 1861; the first trip along the entire length of the line took place in 1862, with soon-to-be Prime Minister William Gladstone on board; and the line officially opened in January 1863. The total cost was £1.3 million. A celebratory banquet was held the day before the official opening, but sadly, Charles Pearson did not live to see it, having died the previous year.

On the first day, about 30,000 passengers travelled on the railway and soon, nearly 12 million passengers a week were using it. The original plan was to run trains to call at alternate stations; to avoid collisions, arrival

A trial trip, taken while excavation work was still taking place.

Baker Street, above and below ground.

Below ground, a steam train pulls into Baker Street Station.

and departure of each train was telegraphed ahead from stations en-route. It was planned for there to be a space of at least one station between trains.

The carriages on some trains were divided into first, second and third class, and from 1874 onwards, designated as smoking and non-smoking. They were lit by gas.

THE INGENIOUS VICTORIANS

Steam engines underground

With steam trains the only practical method of pulling the carriages at the time, a method needed to be found that did not involve fumes from engines suffocating passengers as the engines ejected steam and sulphurous smoke in the way they did above ground. One solution that was considered involved an engine that ran in the conventional way above the ground, but which on entering a tunnel, switched to stored heat from pre-heated bricks. The idea was a failure, so new condensing engines were developed, in which steam was diverted from the usual exhaust system, to be condensed and delivered back into the water tanks, and thus emitted less steam. That which did need to be exhausted was discharged when the trains emerged briefly into the open before plunging back into the tunnels.

At Leinster Gardens in West London, two houses were demolished, and then new fake facades built to link the terrace on each side to give the impression of one continuous row of houses. The space behind them was left empty and uncovered as a place for the trains to vent their steam. The fake houses remain today, identifiable from their neighbours only by their blacked-out windows.

From the front and back: The fake houses in London's Leinster Gardens.

TRAINS IN DRAINS

Even so, the smell of the air below ground could still be foul. In an attempt to put a positive spin on the situation, railway owners suggested the atmosphere was actually invigorating and would be a healthy boon to those who suffered from asthma. Train drivers were permitted to grow beards in an attempt to filter out the fumes.

New lines
Following the success of the Metropolitan railway, various entrepreneurs began investing in the building of new lines, first the District railway along the Victoria embankment and then the Circle line that linked it back to the Metropolitan. By then, the cut and cover method had been abandoned in favour of tunnelling directly through the ground, using a shield, similar to the one used by Brunel nearly sixty years before when digging the Thames Tunnel.

The shield used for the London Underground construction was circular, rather than rectangular as used by Brunel. It consisted of an iron

The shield used to bore tunnels after the cut and cover method was abandoned.

frame with compartments, each of which contained men with spades. They were drawn from navvies who had been involved in building Britain's canal system, and from Cornwall's tin mines. The diggers were required to dig short distances into the ground, and then timbers were erected to support the earth as hydraulic jacks pushed the frame forward and bricklayers followed to build a brick wall around the tunnel that had been created. They were all paid by the yard.

Building of other lines continued in a fairly haphazard way, due to so many different private companies getting involved without there being any kind of central plan for the whole infrastructure. Very often, companies operating one line refused to sell tickets for their rivals' lines, which led to many inconsistencies and complications for passengers who, in order to change trains, sometimes had to leave one, return to street level and then enter another via a new entrance. Different companies built different styles of trains, some even without windows because it was considered that there was nothing for passengers to see outside. Passengers, who thought it was like travelling in padded cells, showed little enthusiasm for that idea. One of the many entrepreneurs who successfully built one of the underground railway lines at this time was Sir Edward Watkin, the man whose previous rather more unsuccessful projects had included the first Channel tunnel and London's answer to the Eiffel Tower.

Steam trains continued to be used on the Underground throughout much of Queen Victoria's reign, although electrification began when the City and South London Railway – the first deep-level line – opened in 1890. From then on, electricity gradually replaced steam. The famous red circle with a horizontal line logo was launched in 1908. The first version of the Underground's iconic map made its appearance in 1933.

Back in 1863, that first Metropolitan line ran for a little over 3 miles. Today, the London Underground covers more like 250 miles and carries about 1.1 billion passengers a year.

CHAPTER 15

How to Avoid Being Buried Alive

In 1844, a short story called *The Premature Burial*, written by American author Edgar Allan Poe, was published. Its plot dealt with a man who was prone to catalepsy, a nervous condition that caused its patient to periodically fall into a rigid, unmovable state that could be mistaken for death. The story's hero is paranoid about falling into such a state when he is away from people who are aware of his condition, being taken for dead and then buried alive. To counteract this, he builds his own tomb in advance of his death, which he equips with various devices that he might use to signal his awakening from a catatonic trance, should the worst happen.

The story, of course, was fiction. But it was based on – and pandered to – taphophobia, or a paranoia of being buried alive that was rife in the nineteenth century. It was a time when medical science was not always efficient enough to tell the difference between death and symptoms that were very similar: hypnotism, being struck by lightning, hysteria and sunstroke, for example. There were a number of cases that proved the fear was not totally without foundation.

Join the club
Such was the age's preoccupation with premature burial, in America, a group was formed that called itself the Society for the Prevention of People Being Buried Alive. In the 1850s, the Society offered, for a small fee, a pre-burial service to give peace of mind to anyone worried about the slightest possibility of premature burial. It included:

1. Examination by a doctor guaranteed to be sober at the time of the examination and, if possible, a dedicated teetotaller.
2. A three-day cooling-off period under constant supervision.
3. A second examination by an alcohol-free doctor.
4. A gunshot to the face after the funeral and prior to burial.

THE INGENIOUS VICTORIANS

5. A coffin with a string inside, leading to an external bell that might be rung by a person if the very worst came to the worst.

In Britain, in 1896, the London Association for the Prevention of Premature Burial was established with the aim of safeguarding its members against being buried alive, as well as for the promotion of effective legislation for the compulsory verification of death.

As the fear of premature burial spread through the Victorian era, businesses were set up to design and build special vaults that would allow the escape of unfortunate victims. Typical of these was one that contained several compartments built into a stone structure, each one with a lid that could be opened from inside by turning a hand wheel.

This Victorian paranoia is perhaps best epitomised by the case of Timothy Clark Smith, a nineteenth-century doctor and lifelong taphophobia sufferer who designed his own safety coffin, which he decreed should be built for him after his death. The coffin incorporated a 6-foot long shaft with a glass window at the top, so that when he was buried, the window would be at ground level and mourners could use it to keep a watch on the body. He also instructed that a bell should be erected with a string leading down the shaft to the coffin.

So paranoid was Dr Smith that he also had a second vault built for his wife, incorporating an underground room from which a flight of stairs led to the surface.

The gravestone of Dr Timothy Clark Smith, with its glass window to allow mourners and other passers-by to inspect the corpse 6ft below.

HOW TO AVOID BEING BURIED ALIVE

When Dr Smith died, appropriately enough on the night of Halloween 1893, his wishes were carried out, and his gravestone with the 14 x 14 inch glass window in its centre can still be seen today at Evergreen Cemetery in Vermont.

Pre-burial safety coffins
Safety coffins in various forms proliferated through the Victorian era, and were largely split into two types: those from which a person might escape prior to actual burial, and those that offered some form of signalling to the outside world after the coffin had been buried.

Inventor Johan Toolen from the Netherlands patented a device that enabled the buried person to open the coffin from the inside. It consisted of a special spring-loaded lid, held in place by a bolt, connected to the person in the coffin. The mechanism was such that the slightest movement of the corpse would disengage the bolt and the lid would atomically spring open, at the risk of inducing a nasty shock to any mourners nearby.

Russian inventor Michael Karnicki suggested a coffin with a glass panel in the lid, which was also sprung so that it could be opened from the inside. If the buried person awoke, he or she would see light coming through the glass window and be able to spring the lid open. The patent anticipated the problem of revival in the hours of darkness when no light might come through the window, but added: 'If the awakening of the sick person takes place at night, it is inevitably necessary to wait for some hours, but with the beginning of day, service at the church-altar will afford a certain chance of speedy aid.'

American George Field had a much simpler idea: simply make the coffin of glass so that those outside could see in if all was not well.

Post-burial safety coffins
Of course, if nothing untoward had taken place between the death (or otherwise) and the coffin being actually buried, then different measures needed to be taken.

In 1822, Dr Adolf Gutsmuth, in an attempt to prove to himself and demonstrate to others that premature burial needn't be fatal, had himself buried in a coffin of his own design. Built into it was a tube to allow air in and through which he could be fed. Lying underground for several hours, he devoured a meal of soup and sausages, washed down with beer, delivered to him via the tube, before being dug up none the worse for his experience.

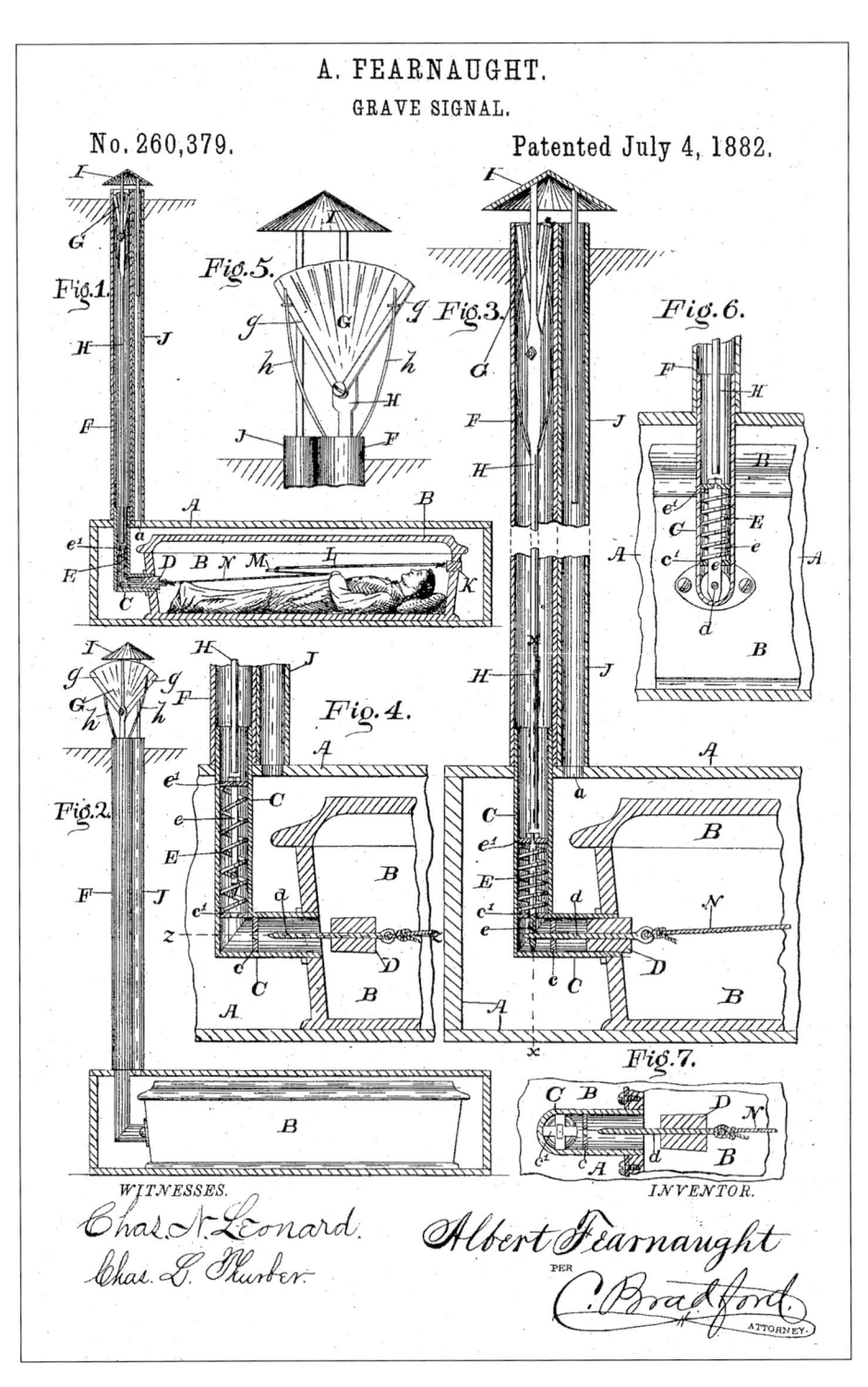

The complicated patent drawing of Albert Fearnaught, whose safety coffin allowed in air and raised a flag when movement was detected.

HOW TO AVOID BEING BURIED ALIVE

In Germany, more than thirty patents were filed for safety coffins during the nineteenth century, epitomised by one from Dr Johann Gottfried Taberger, in which the corpse's hands, feet and head were attached by strings to an external bell. The thinking was that any movement of the person buried below ground would activate the bell above ground. It formed the basis of a design suggested in different forms

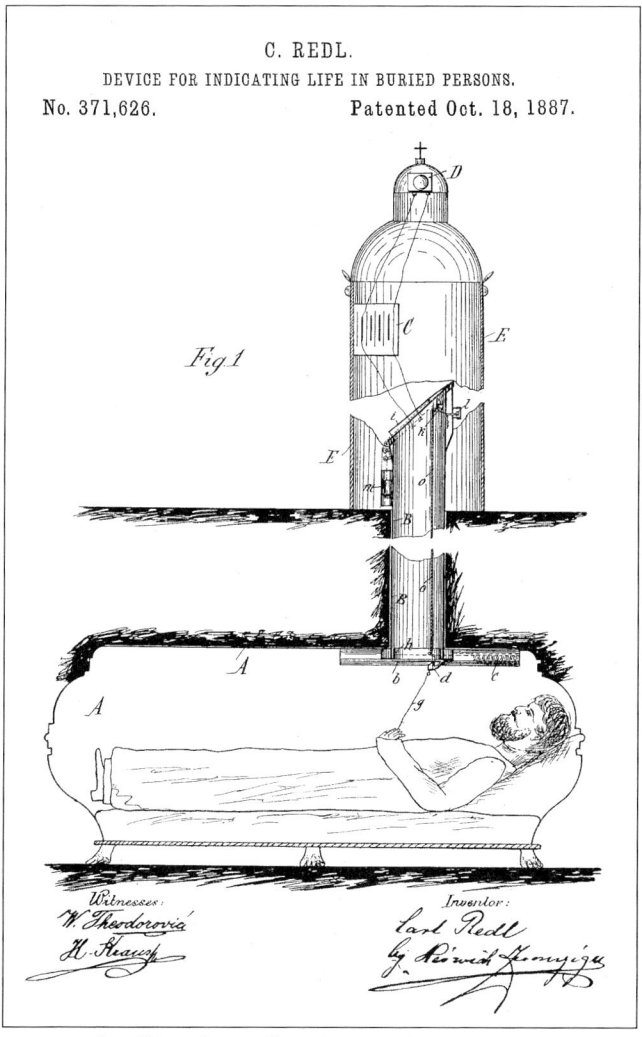

Vertical section of Redl's safety coffin, showing the air pipe, the plates that could be opened and closed, and the electrical connection to the bell.

155

by various inventors, but had a design fault that most had overlooked. As a body decomposes, it bloats, causing it to move slightly within the coffin, actions that inevitably set the bells ringing several days after death and burial, much to the to the amazement – and no doubt the horror – of anyone in the vicinity of the grave.

An electric alarm bell was employed by Viennese inventor Carl Redl, whose design incorporated batteries for the purpose, in a type of coffin that could be modified, according to whether the person was buried in the ground or placed in a vault. In the first version, an air pipe was placed over a hinged plate that could be opened by the buried person in order to get fresh air into the coffin. As the plate was opened, an electric circuit was closed to trigger the bell, situated either on the grave or at a place nearby. A similar version with a shorter air pipe could be used if the coffin was placed in a vault or sarcophagus. His version also activated an electric bell if the air pipe was uncovered.

American inventor John Krichbaum designed a coffin in which a T-shaped pipe passed down a shaft from the surface to the coffin and then through the lid to rest on the body near the hands. Any movement from within that caused the T-bar to be twisted would be indicated on a scale outside. Further, more violent movement would cause a cover over the pipe to move aside and let in air, sufficient to keep the person alive until help arrived. The mechanism was so made that the pipe cover could not then be closed again.

Americans Franz Egerland and John Freese suggested a coffin of fairly conventional design, but one that incorporated a tube or pipe that led from the surface of the ground and into the coffin. Halfway along the tube, however, they introduced a container of disinfectant to prevent the discharge of poisonous or obnoxious gases from the coffin. The pipe also carried wires from a switch and battery inside the coffin to an electric bell on the surface. The switch was said to be activated by the slightest movement of the corpse.

It is sometimes stated that designs of this type were responsible for the expression 'saved by the bell', but it is more likely that this phrase originated later in the world of boxing, where ringing the bell at the end of a round might save one of the fighters from defeat. Similarly, the expression 'dead ringer' was also wrongly associated with graveside bells; it comes from horse racing, in which a horse is fraudulently substituted for another in a race.

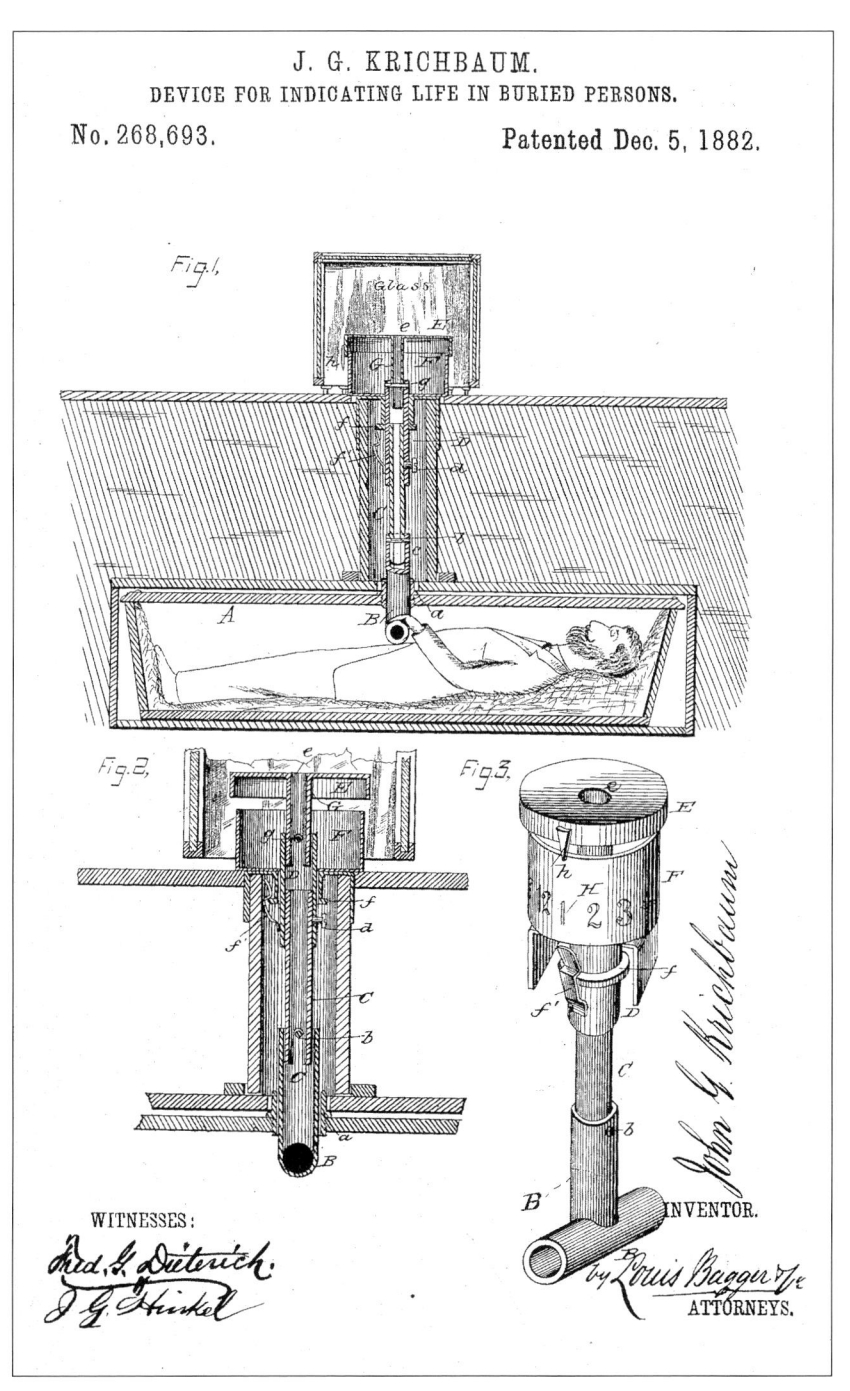

John Krichbaum's safety coffin, whose design incorporated a T-bar resting on the corpse in order to detect movement.

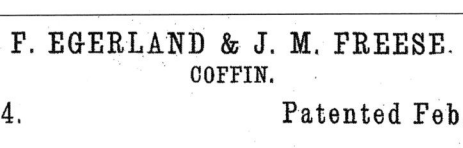

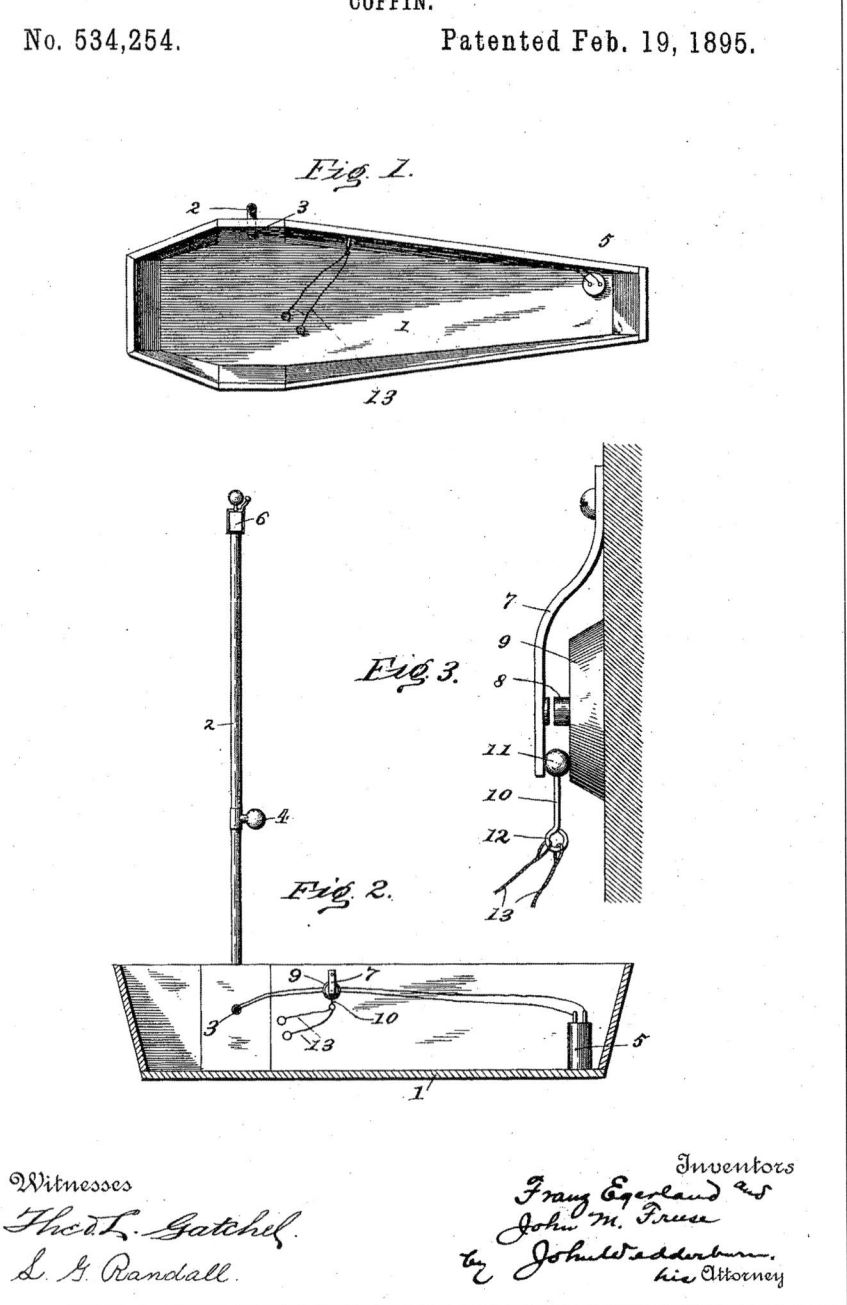

Two views of the coffin and the bell-push device that was designed to detect movement and ring an electric bell.

Advertisement from American safety coffin makers Bobeth & Schulenberg.

Other designs

Numerous other safety coffin designs suggested ways in which movements within coffins could activate mechanisms, often coupled with a window in the lid that could be viewed by the outside world. Among the many ideas patented and made were devices that opened vents to allow air in; battery powered lamps within the coffin that would come on when sensing movement, which might then be seen through the coffin's glass window; a spring-loaded hammer that would fall to break the window glass above the corpse, to allow in air; a flag, made to spring to attention when movement beneath the surface was detected; a method of sending a telegraphic message to the home of the cemetery administrator or local police station; and even a way to set off a firework rocket.

Bobeth & Schulenberg of Baltimore were among the many who produced a product in which springs and levers on the inside ensured that movement would cause the coffin lid to fly open. In advertising their coffins for use 'in doubtful cases of death', they offered clear instructions to how families might use them:

> The inventor would advise all families or communities who may use his Life-Preserving Coffins, to place them in a room or in a vault with a lock on its door that will open on the inside or outside, the key which should be placed in the inside to allow the individual (who may be resuscitated) egress, or to remain until decomposition takes place, when they may be removed and finally deposited in the ground.
>
> Whenever it can be done, the inventor suggests that the coffin be kept in a properly heated room which, from the construction of the apparatus, obviates all danger from without, at the same time admitting freely pure air within, and with a string attached to the hand of the inmate, leading to a bell which will, in case of resuscitation, instantly give the alarm, and provide all necessary attention.

The advert contained many testimonials from people who had examined the manufacturer's coffins and proclaimed them things of beauty and ingenuity. There were no testimonials from anyone who had had the misfortune to inhabit one of them.

CHAPTER 16

London's Great Stink

In the early 1800s, the banks of the river Thames in London were pleasant areas to live, work and visit. Fish thrived and could be caught in the waters between the Vauxhall and London bridges; in the upper reaches it was not uncommon to find salmon; and the area supported a thriving community of fishermen. The waters of the Thames were clear, pleasure boats abounded and the air was fresh. Houses whose gardens reached down to the river were highly prized and inhabited by the wealthy. The area provided Londoners with an attractive place to meet and walk on a summer's evening.

By the middle of the century it was a different story. The high-class houses had been abandoned and were growing derelict, the river was filthy and stinking, and the fish were dead and gone. No one took to the water unless they were forced to do so for commercial purposes.

The reason for such a disastrous change can be summed up in one word: sewage.

As the first flush toilets began to appear, the amount of water and waste that poured into cesspits increased, frequently overflowing into the drains and elementary sewers, most of which were little more than open ditches sloping down to the Thames. The drainage system had originally been designed to cope only with rainwater, but in the 1840s, cesspits were abolished as it became compulsory to drain household waste into sewers. Soon the overloaded and overflowing drainage system not only took the sewage from the houses, but also the outflows from factories and slaughterhouses. By then, the population of the city that had grown up along the banks of the river amounted to about 2.5 million people who, between them, produced some 60 million gallons of sewage and general filth every day, all of which oozed its way along the streets' inadequate drainage systems to be dumped straight into the Thames. To compound the situation, Londoners were prone to throwing the contents of their chamber pots out of upstairs windows, into the streets below.

The streets of London smelt vile and the river was little more than an open sewer. To make matters worse, there was no real central body to oversee the drainage and sewage systems of London, just eight different districts with their own commissioners, who were concerned only with their own areas.

One factor that led to the creation of a proper sewage system for London was the discovery that cholera, of which there had been several outbreaks since the first case had been recorded in 1831, was not spread in the way many thought it to be. The popular belief at the time was that the disease was spread by bad smells in the air. But deduction by physician Dr John Snow traced one major outbreak to cross-contamination of water supplies and the inadequate sewage system.

Things came to a head in the particularly hot summer of 1858, when a combination of the heat and the heavily polluted Thames resulted in Members of Parliament finding it necessary to hang sacking soaked with deodorising chemicals at the windows of the House of Commons that

The mills at Plaistow in East London, where concrete was made for the sewers.

Laying the concrete foundations of the Northern Outfall tunnels.

Constructing the concrete embankment across the Plaistow Marshes.

bordered the Thames at Westminster. It became known as the year of The Great Stink.

The summer ended with heavy rain that helped clear some of the problem, but by then the government, prompted by the smell literally outside their own windows, had rushed a Bill through Parliament to provide the money to construct an ambitious sewer system for London. The Bill became law in eighteen days, and the man who took on the task of designing the system was Joseph Bazalgette, Chief Engineer of the Metropolitan Board of Works. Work began in 1859 and carried on for nine years, at a cost of £4.2 million. It became the biggest civil engineering project in the world.

THE INGENIOUS VICTORIANS

The main objects of the new system were waste disposal, land drainage and the introduction of safe drinking water. It had also to take into account that the river Thames was tidal. Therefore, anything discharged into the river in Central London might be washed downriver with the tide, but would surely return when the tide turned. To be effective, if waste were to be dumped in the river, it had to be at a point closer to the Thames Estuary, and from there into the sea.

Pumping Station at Deptford.

Driving a tunnel at Peckham.

LONDON'S GREAT STINK

The proposal was to intercept the sewage on its way to the river, and by means of both gravity and pumping stations along the way, divert it to two locations in the Thames Estuary, about 14 miles east of London Bridge and at a suitable distance from the main population of London. The northern outfall site was at Barking Creek, a few miles outside the main town of Barking, in Essex. The southern outfall was at Erith Marshes, which at that time stretched between the towns of Erith and Woolwich, in Kent.

At each of these places the sewage was contained until the commencement of the daily ebb tide, the period between high tide and low tide when the water of the river flowed towards the sea. During the first two hours of this period, the sewage was pumped into the bottom of the river, where it would be diluted by volumes of water twenty times greater than was the case in Central London, whilst conveying it another 12 miles towards the sea, never to return.

The construction of the sewers was complicated by the fact that London was built on many different levels, ranging from the high elevations of places like Hampstead and Highgate in the north to districts like Lambeth in the south, much of which was actually below the level of the river. A good deal of the system was built underground, but in places the drainage and the pipe systems involved were routed above ground, over rivers, canals, railways and roads, involving engineering work to lift the outflow from low to higher levels. Work also had to be carried out alongside the London Underground system, which was in development at the same time.

Outfall Works at Barking Creek.

Bottom of a shaft in the southern high-level sewer at Peckham.

The system on the north side of the Thames was divided into three distinct drainage areas, called the high, middle and low areas, each separated by a main sewer, cutting off at right angles all the local drains that ran into the Thames. The three met at what was called a penstock chamber, the name given to a gateway designed to regulate the flow of waste and water. From there, the sewage was pumped onwards to the outflow at Barking Creek.

The southern system worked in a similar way, but with only two drainage areas instead of three.

In an account such as this, it would be pointless (not to say a little boring) to go into intricate details of the construction of every branch of the system. But, to give an idea of what was involved in the building, and the materials used, consider just one small part of the whole network.

The northern high-level sewer was 9 miles long from its start in Hampstead to the penstock chamber at Old Ford. From Hampstead, it passed through Stoke Newington, intercepting the old river Fleet sewer, which would otherwise have emptied its contents into the Thames at Blackfriars Bridge, as well as the old Hackney Brook drain. On its course,

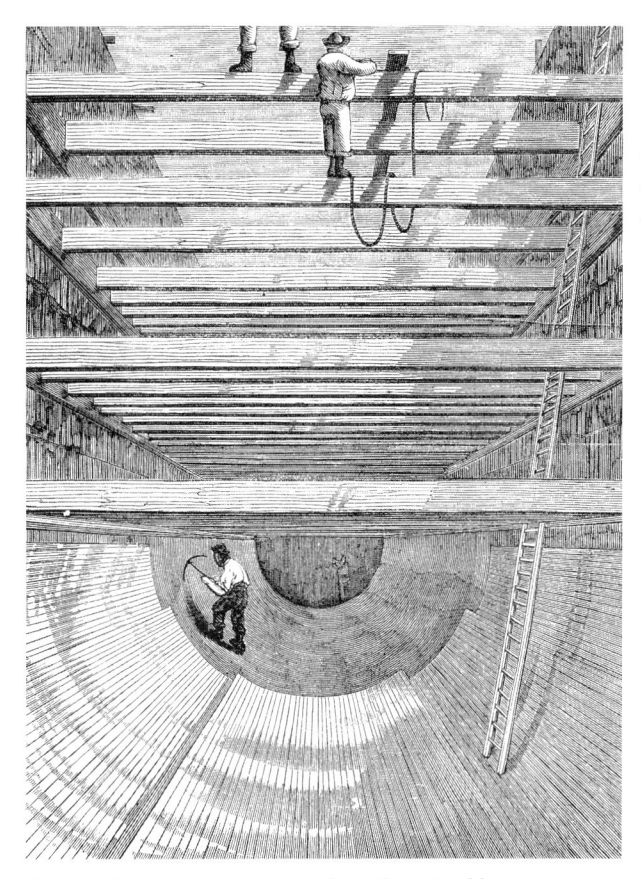

Construction of the high-level sewer at Peckham.

Barrow hoist to lift excavated earth at Peckham.

Construction of the Fleet sewer.

The penstock chamber at Old Ford.

it passed under the Great Northern Railway and the New River, an artificial waterway opened in 1613 to supply London with fresh drinking water, then on to Old Ford. At its upper end, the sewer was 3 feet in diameter, growing to 12 feet, as subsidiary sewers were connected to it, when it reached the penstock chamber.

This section alone involved excavating about 1 million yards of earth from the trench in which it was constructed, and the laying of 40 million bricks. To form the foundations, backings and coverings, 100,000 cubic yards of concrete were deposited, involving 200,000 bushels of Portland cement and 350,000 bushels of lime. Among other materials used were 20 tons of hoop iron. The average depth of the sewer varied from 30 to

THE INGENIOUS VICTORIANS

Memorial to Joseph Bazalgette on the north bank of the river Thames in London, close to Hungerford Bridge.

LONDON'S GREAT STINK

50 feet and it was built on a slope that drove its contents along at about 3 miles per hour.

In November 1861, while building was still in progress, writing about the southern system, *The Illustrated London News* summed up the achievements of all concerned thus:

> The works already executed throughout the whole length of the southern mains drainage are of unexampled excellence, a most unusual and, in the eyes of contractors, most unnecessary degrees of perfection being assisted upon, by Mr Grant [John Grant, Assistant-Engineer to the Metropolitan Board of Works], under whose personal superintendants the whole is carried out, at all points deep in the belt of the earth, where no human being will ever see it after it is once in work. The brickwork is executed as beautifully as though it could be handled in the most aristocratic thoroughfares of the metropolis. This is as it should be. These important works that, when once finished, are buried for ever, require to be much greater excellence than those above ground, whose defects, if any, maybe easily discovered and remedied.

When Joseph Bazalgette began building the system, it was to serve a population of about 2.5 million, and he anticipated that it would cope with a population that could rise to 4 million. Today, an extended version of that same basic system serves more like 8 million Londoners.

CHAPTER 17

How London Got the Needle

Cleopatra's Needle, which stands today on the north bank of the river Thames, dates back more than 3,500 years, with its origins in Ancient Egypt. But without the entrepreneurial vision of one Victorian, the engineering genius of two others and the bravery of several more it wouldn't be where it is today. Neither would it be flanked by its two gigantic sphinxes, which, far from being Egyptian and ancient, were actually designed and built by the Victorians. It is perhaps also worth mentioning that it has little to do with Cleopatra.

A brief history
About 1,500 years before the birth of Christ, and more than 1,400 years before Queen Cleopatra VII (with whose name the Needle has been inextricably linked) was on the throne, the obelisk was carved and built for Thutmose III, the sixth pharaoh of the Egyptian 18th Dynasty.

The Needle was made from red granite from the quarries of Aswan and stood approximately 69 feet high with a weight of more than 200 tons. In about 1500 BC it was erected at the entrance to a temple in Heliopolis, and there it stood for nearly 1,500 years. Then, in 12 BC, eighteen years after the death of Cleopatra, the Romans, who were by now ruling over Egypt, shifted it to Alexandria, where it was erected in the Caesareum, a temple that had been built by Cleopatra in honour of Julius Caesar and Mark Antony, both of whom had been her lovers. This is the only tangible connection between the monument and the name of the Egyptian queen.

More than 1,800 years later, in 1798, while anchored off the coast of Alexandria during the Battle of the Nile, Napoleon's fleet was ambushed and defeated by English warships led by Horatio Nelson. Plans were made to bring the Needle to Britain in 1801, as a memorial to Nelson's victories, but that didn't happen. In 1820, in recognition of not just Nelson, but also

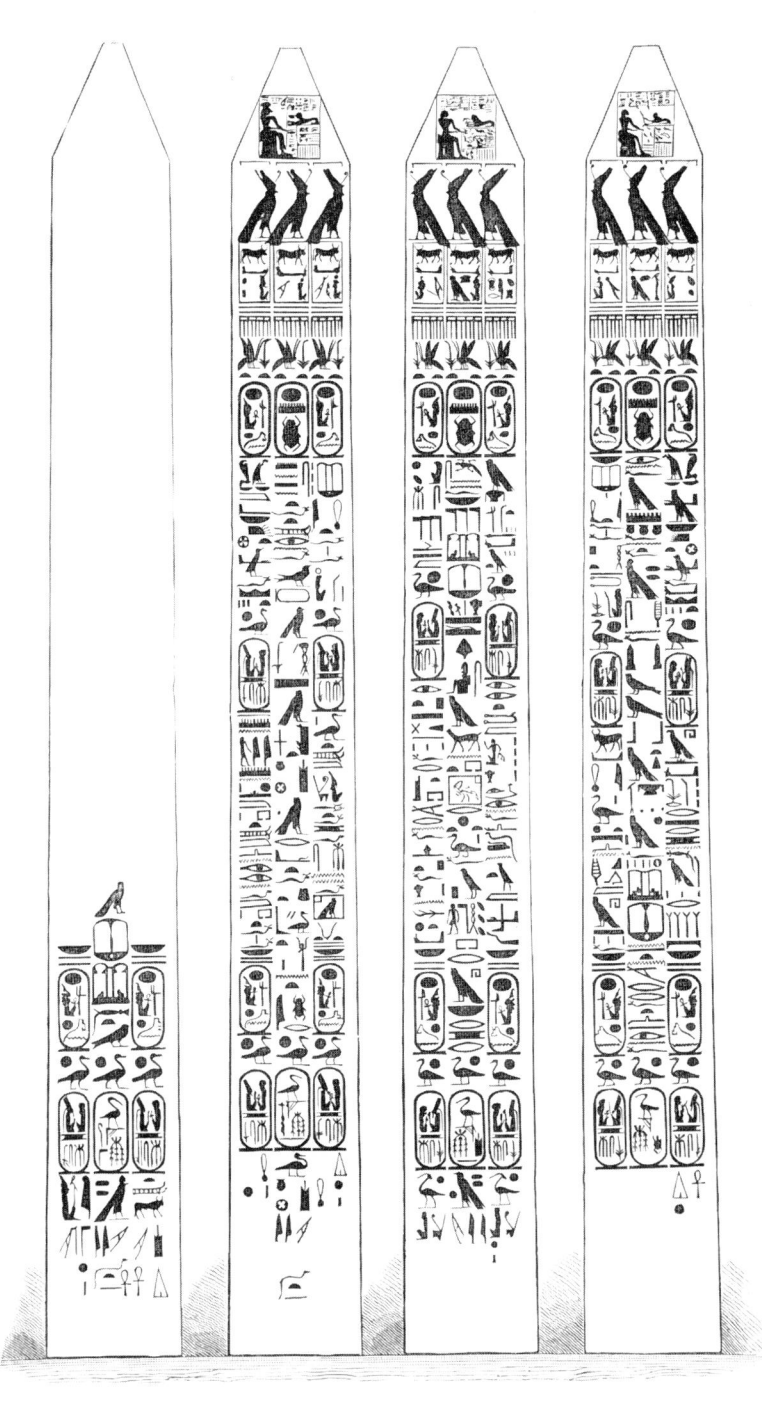

The hieroglyphics on the four faces of the Needle were in honour of Thutmose III and Ramesses II.

the success of British troops under Sir Ralph Abercrombie during the Battle of Alexandria, Egyptian ruler Sudan Muhammad Ali presented the monument to the British nation.

For the British Government, this represented good news and bad news. The good news was that Britain had been recognised and honoured in this way by Egypt. The bad news was that although the Egyptians were happy for Britain to take ownership of the Needle, they didn't actually make preparations for it to be transported to Britain. British Prime Minister Robert Banks Jenkinson was quick to decree that there were not enough resources for Britain to pay for transportation, and so Cleopatra's Needle, although officially British owned, remained in Alexandria for close to sixty more years, by which time Queen Victoria was on the throne.

Enter, at this point, Sir William James Erasmus Wilson. He was an English surgeon and dermatologist who had made a huge fortune from his medical skills and shrewd investments. He was also a great philanthropist who devoted much of his money to worthwhile charities and educational projects. In 1877, he put up the cash to have Cleopatra's Needle transported to Britain.

The voyage to London
The task of transporting a nearly 70-foot long length of granite weighing about 200 tons fell to railway and civil engineer John Dixon, who accepted the challenge of being paid £10,000 if the task was completed successfully – and nothing at all if he failed. Dixon commissioned fellow engineer Benjamin Baker to design a method of transportation.

In Alexandria the Needle was first encased in a wrought iron cylinder, designed by Baker. Using a system of levers and chains, the cylinder, with the Needle inside, was then rolled to the beach at Alexandria, where it was fitted with a deck, mast, rudder and steering gear, effectively transforming it into an ocean-going vessel, and given the name *Cleopatra*.

Manned by Maltese sailors, it set off for Britain in September 1877, towed behind a steamship called the *Olga*, each vessel with its own captain and crew. As the *Cleopatra* lay low in the water, waves continually washed across the prow, but Baker's design ensured that the waves were split as they progressed along the cylinder, and the deckhouse remained dry.

Progress was slow at about 7 knots, but in October, about 2,500 miles

The Cleopatra *on its way to Britain, towed by the* Olga.

into the journey, disaster struck as the pair of vessels crossed the notoriously stormy Bay of Biscay. Buffeted by a heavy storm, ballast in the *Cleopatra* shifted and the ship rolled onto its side, the deckhouse disappearing beneath the waves.

The captain of the *Olga* was forced to give the order to cut the tow ropes, and a rescue mission was launched to get the captain and eight-man crew off the *Cleopatra*. Six men from the *Olga* lost their lives during the attempt when their boat was swamped.

When the sea calmed, the *Olga* managed to move alongside the *Cleopatra* and get two lines across between the vessels. A boat was lowered, and the *Cleopatra's* crew were rescued and taken on board the *Olga*. The *Cleopatra* was then cast adrift.

Later that day, the iron cylinder, still with the Needle encased within it, was spotted by the crew of the Glaswegian steamer *Fitzmaurice,* who managed to get aboard to carry out minor repairs and then tow it into the harbour at Ferrol in Spain, where further repairs were undertaken. From there, the Needle was towed again, this time by a paddle tug called the *Anglia*, specially built in England and sent to Spain for the purpose. Four months after leaving Egypt, it eventually hit English soil at Gravesend in Kent, in January 1878, where it received a rapturous welcome from crowds who had been following its progress in the English newspapers and magazines like *The Graphic* and *The Illustrated London News*.

THE INGENIOUS VICTORIANS

Raising the Needle

Erasmus Wilson favoured Parliament Square as a location for the Needle and a wooden model was erected there to give an idea of what it might look like. The final decision, however, was to erect it on the Thames Embankment. Not everyone was in favour of the chosen location. *The Engineer* magazine of 1 March 1878 had this to say:

> We regret that in defiance of good taste, it has been decided to erect the obelisk on the Thames Embankment. We have spoken so frequently and so strongly on the selection of the site that we shall not further refer to the subject now. It is well known, however, that no objection can be urged on engineering grounds to the position secured on the Embankment. In other words, there is no fear that the monolith will tumble into the river, bringing a large section of the Embankment with it.

The machinery for placing the obelisk into position on the Thames Embankment.

HOW LONDON GOT THE NEEDLE

As the Needle made its final trip from Gravesend, along the Thames and into the heart of London, a huge wooden structure, also designed by Benjamin Baker, was erected at the side of the river. It comprised four 50-foot high, diagonally braced posts, at the base of which the obelisk, stripped from its iron cylinder, was inserted.

Hydraulic jacks were used to raise the horizontal monument to a point where it could be swung through 90 degrees to an upright position and lowered onto its base. In Victorian times, the trees along the Embankment were mere saplings, with the Needle towering impressively above them. Today, the trees are as high as, if not higher than, the obelisk itself.

Cleopatra's Needle in Victorian times, dwarfing the trees along the Embankment.

One of the two Victorian sphinxes that flank the Needle.

The arm rests of seats along the Embankment close to the Needle are made in the form of winged sphinxes.

HOW LONDON GOT THE NEEDLE

This was at a time when Londoners were showing a great deal of interest in Ancient Egypt. The grounds of the Crystal Palace, which had been rebuilt in South London in 1854, contained large stone sphinxes at the bottom of several staircases. A road in Islington contained houses guarded by miniature sphinxes and obelisks. Highgate Cemetery featured an Egyptian Avenue. And of course, the British Museum was full of Egyptology.

So it's not surprising to find that the Victorians decided to build their own sphinxes to stand each side of Cleopatra's Needle. They were designed by English architect George John Vulliamy, who was also responsible for the bronze pedestal on which the Needle stood, as well as several London buildings, fire stations and lamp posts with dolphins entwined around them along the Embankment. He also designed benches adjacent to the Needle whose armrests resembled winged sphinxes.

The huge sphinxes on either side of the Needle were cast in bronze with hieroglyphic inscriptions added that translated as 'the good god, Thuthmosis III given life'. The sphinxes' original purpose was to stand guard over the Needle, in which case, they should be facing outwards and away from the monument. In fact, both face the Needle, since Queen Victoria thought that position to be more aesthetically pleasing.

Cleoptra's Needle today.

CHAPTER 18

Phonographs, Graphophones and Gramophones

Thomas Edison was one of the great inventors of the Victorian age who, in his lifetime, held more than 1,000 patents. Yet many of the inventions with which he is today credited were not completely original. His great talent was to take an existing idea, then apply his own thinking in ways that vastly improved on the concept and led to its commercial success.

It is sometimes suggested, for example, that he invented the electric light bulb, when in fact many inventors before him had shown ways of making wires glow to produce elementary forms of electric lamps. What Edison did was to find a way to greatly extend the short life of these early light bulbs and so make their use far more practical. It led, by 1878, to his founding of the Edison Electric Light Company, giving the first demonstration of his incandescent light bulb the following year.

Inventor and businessman Thomas Edison.

Thomas Alva Edison was born in 1847. He was a hyperactive child whose incessantly enquiring mind was often the despair of his teachers. When he was eleven years old, in an attempt to quench his thirst for knowledge, his father introduced him to the local library, where he read everything he could get his hands on, including a great many scientific journals.

At the age of twelve, he started his own business, selling newspapers and snacks to railway commuters. Then, using news releases being teletyped into the railway station, he began his own newspaper to sell to commuters. It became the first newspaper to be typeset, printed and sold on a moving train. He learnt Morse code and, at fifteen, took a job as a telegraph operator, a career he pursued with several employers and which led eventually to his interest in the possibilities of recording and playing back sound.

As far as sound recording was concerned, at least two inventions preceded Edison's. First there was the phonautograph, patented in 1857 by its inventor, French printer and bookseller Édouard-Léon Scott de Martinville. His invention recorded sound waves, and then displayed their vibrations visually on paper coated with soot. The result could be observed but not played back audibly. In 1877, French poet and amateur scientist Charles Cros suggested the paleophone, which proposed the use of photoengraving, a process already used in the production of printing plates, to convert the soot tracings of the phonautograph into grooves and ridges on a metal cylinder. With a diaphragm linked to a stylus that sat in the groves of the cylinder as it revolved, the original sound would be reproduced.

Edison's phonograph
It was Edison, however, who made all previous experiments more practical and commercial by producing the first machine to both record and play back sound. Of all his inventions, it was said to have been his favourite. In his patent filed in 1877 and published in 1878, he called it his Phonograph or Speaking Machine. The patent stated:

> The object of this invention is to record in permanent characters the human voice and other sounds, from which characters such sounds may be reproduced and rendered audible again at a future time.

After a long series of experiments, Edison had discovered that a diaphragm when set in motion by sound did not, as had been supposed, produce a constant vibration. Instead, it produced many vibrations, each of which was separate and distinct according to the sound. From this he deduced that it would be possible to use a stylus attached to the diaphragm to record and then play back an audible reproduction of the human voice. He first saw this as a useful way to record telegraph messages and play

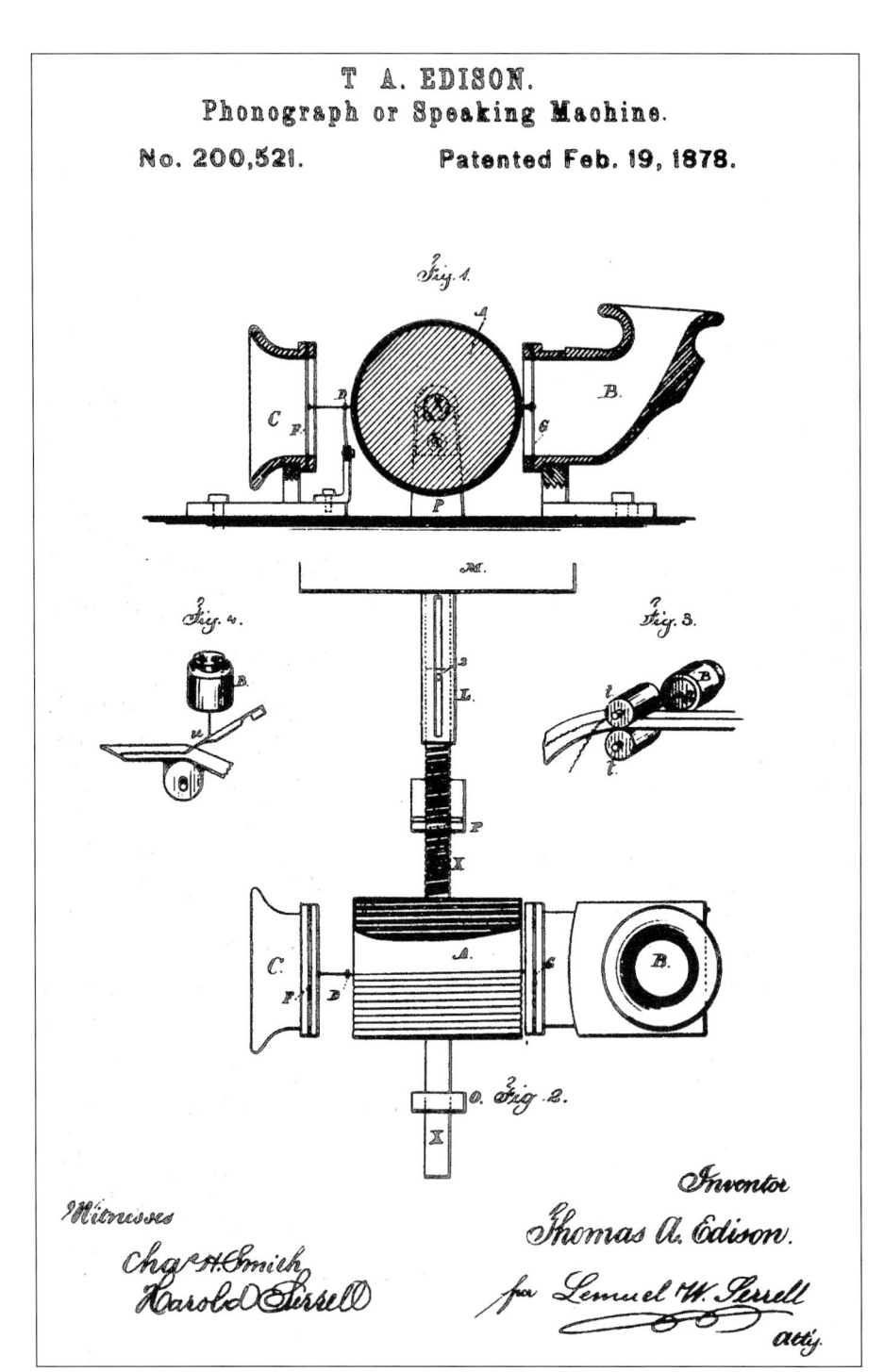

From Edison's phonograph patent of 1878.

them back by telephone, pre-empting the answering machine by a great many years.

His initial attempts utilised a disc of paper on a metal platen, a foretaste of what would come later with gramophone records. Edison, however, did not pursue that route, because he shied away from the fact that the recording grooves on a flat disc would have to take the shape of a spiral, getting smaller as the stylus approached the centre of the disc.

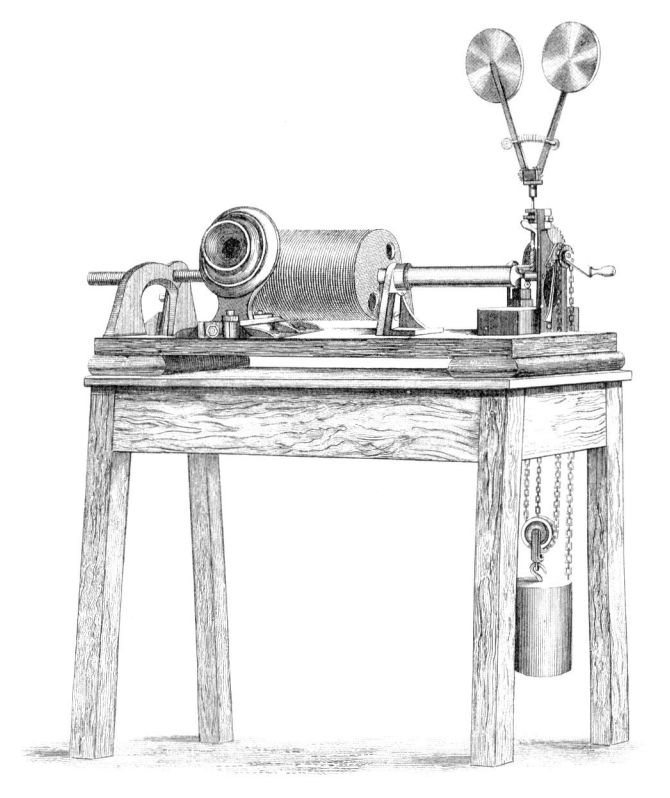

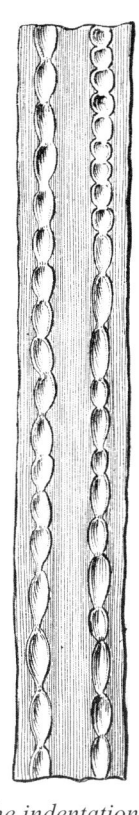

An early version of Edison's phonograph.

The indentations in the tin foil whose patterns reproduced sound.

Instead, he turned to using a grooved cylinder, around which he wound tin foil. He thought this more practical, since the grooves remained uniformly the same size across the length of the cylinder, along which the stylus attached to the diaphragm travelled. In early versions of the machine, the cylinder was revolved by a falling weight mechanism and spinning paddles that regulated the speed. A later improved version used battery power to drive a small electric motor.

As the cylinder revolved, indentations in the tin foil as the stylus sunk into the cylinder's grooves translated the movements of the diaphragm activated by the sound of a voice. The pattern so produced on the tin foil could then be played back to replicate the sound of the voice. The first words recorded by Edison for demonstration purposes were 'Mary had a little lamb'.

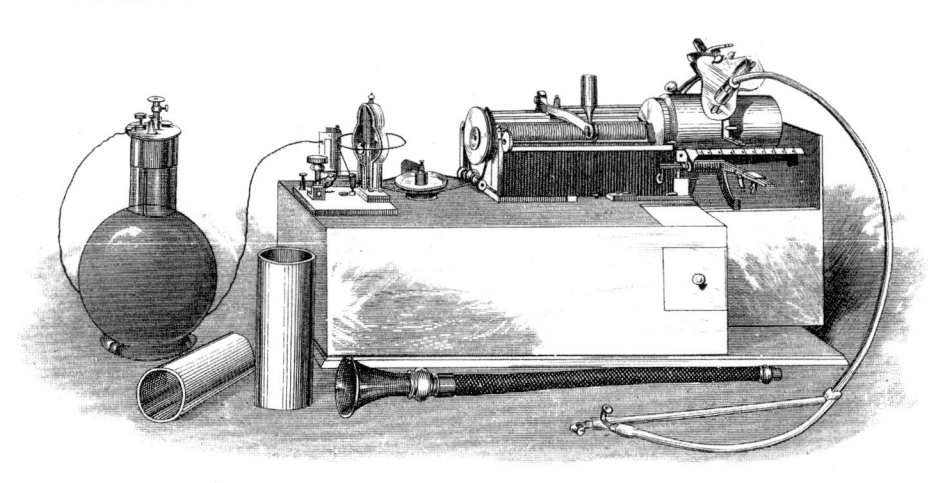

An improved version of Edison's phonograph.

Edison admitted that he was actually surprised how well it worked first time, but went on to improve on his original designs. In 1881, he demonstrated his latest model to music critic Herman Klein, who had this to say about what he heard:

> It sounded to my ear like someone singing about half a mile away, or talking at the other end of a big hall; but the effect was rather pleasant, save for a peculiar nasal quality wholly due to the mechanism.

Edison's phonograph was an amazing achievement for its time, but it was far from perfect or efficient. Rather than improve upon it at that time, however, Edison moved on to other things and it fell to the Alexander Graham Bell Company to take the idea to its next stage.

The problem was with the tin foil used as a recording medium. The quality of the sound it produced was not very good, and the foil tore easily, making it suitable only for use a few times after the recording had been made. The result was that, once the initial novelty of sound recording had worn off, people were slow to buy a machine that, although revolutionary, was not very efficient.

In his 1878 patent, Edison suggested that copies of his cylinders could be made by first pressing the tin foil into wet plaster of Paris, and when set, using this as a mould to make further tin foil copies. 'This is valuable when musical compositions are required for numerous machines,' he pointed out.

The Graphophone
Recognizing that the weakness lay in the use of tin foil, Bell's company formulated a system in which wax was inserted into the grooves in the cylinder. Thus the patterns produced by the stylus attached to the

The graphophone, which used wax cylinders.

Phonograph and Graphophone wax cylinders.

diaphragm were engraved into the far more permanent wax to make a more efficient recording and playing machine. The name Bell gave to his version of Edison's phonograph was the Graphophone. It was driven by a clockwork motor.

It was Bell's Graphophone, then, rather than Edison's phonograph, that first enjoyed financial success, as patents were taken out, and the American Graphophone Company was formed in 1887 to produce and sell the machines. Later, Edison returned to his invention and made his own improvements as, once again, it was his name that became more associated with the machine. As production increased and prices dropped, Edison's phonograph soon became a popular fixture in Victorian homes as a means of playing music and sometimes short comedy routines.

Meanwhile, other experimenters had been working on Edison's initial idea of recording on a disc rather than a cylinder. An idea was proposed,

PHONOGRAPHS, GRAPHOPHONES AND GRAMOPHONES

though never taken further, by the aforementioned Charles Cros, and it was a variant of this that was patented by German-born American inventor Emil Berliner in 1887. In Berliner's design, a diaphragm was linked to a recording stylus that vibrated laterally, rather than up and down as in the cylinder method of recording, while tracing a spiral pattern on a zinc disc coated with beeswax. Immersing the disc in a bath of chromic acid left a groove in the disc where the stylus had removed the wax coating. He called his machine the Gramophone.

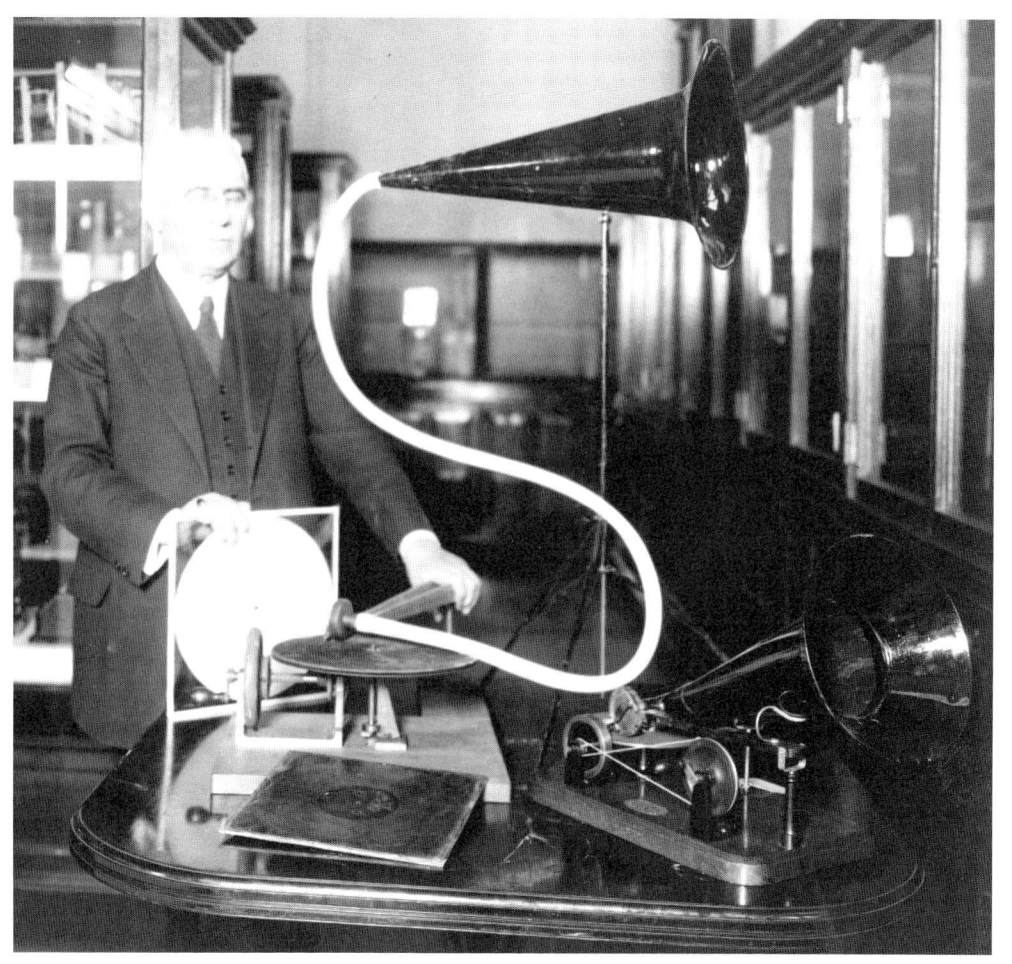

Emil Berliner with his disc recording apparatus.

THE INGENIOUS VICTORIANS

The craze soon spread, and by 1890, record manufacturers were finding fairly basic ways of duplicating records, a vast improvement on wax cylinders, which had to be custom made. Although it had previously been possible to make first just a few and then anything up to 150 cylinders at a time, it still meant that recording artists of the day were obliged to record their performances over and over for commercial purposes.

With the possibility of duplicating records more easily from a master disc, it was the flat disc rather than the cylinder that eventually won the day, and so the record industry began.

An early gramophone, which popularised discs rather than cylinders.

CHAPTER 19

Who Really Invented the Cinema?

The origins of the cinema go back to the Victorian age. Who was the first to invent the process that enables us to watch moving pictures on a screen, however, is a matter of some debate.

In London's Highgate Cemetery lies the grave of portrait photographer and inventor William Friese-Greene. Carved in stone at the base of his memorial are the words 'The Inventor of Kinematography'. To seemingly corroborate that claim, there's a plaque on the wall of a small house in Brighton that proclaims: 'It was here that the inventor of cinematography William Friese-Greene, 1855–1921, carried out his original experiments which led to a world-wide industry.'

So it would seem that the cinema was invented by an Englishman. However, the man more usually credited with producing the first movie camera was American inventor Thomas Edison.

The plaque in Brighton that proclaims William Friese-Greene as the inventor of cinematography.

So where does that leave French inventor Louis Le Prince, whose early work in filming moving images resulted in him being heralded as the Father of Cinematography?

Or American inventor Birt Acres, who, while living in England, claimed he had invented a similar machine to Edison's, three years before Edison made his announcement?

Or Englishman Wordsworth Donisthorpe, who shot moving images of Trafalgar Square in Victorian London, on film, a fragment of which still remains today?

THE INGENIOUS VICTORIANS

The fact is that all of these and more contributed to the invention of cinematography, each one coming up with a different method of making the process work, all of which coalesced into the cinema as we know it today.

William Friese-Greene
William Friese-Greene was a major pioneer of cinematography, whose initial experiments had been to produce a sequence of pictures on long strips of paper soaked in castor oil to make them transparent for projection purposes in a crude projector of his own design. In 1885, he submitted his invention to the Royal Photographic Society, but little interest was shown, possibly because he was seen as not much more than an impoverished inventor with no real pedigree.

Following minor successes with this rather crude system, Friese-Greene moved on to arranging his images on strips of celluloid, which he punched

Cinematography pioneer William Friese-Greene.

with small holes. This allowed the film to be moved through his apparatus and stopped for a fraction of second at each frame. He had the good sense to patent that idea, and its descendants remained the standard for many years to come. Here's what the *Optical Magic Lantern Journal and Photographic Enlarger* publication said about his invention in 1891:

> Mr Friese-Greene has invented a peculiar kind of camera, to outward appearances not unlike an American organette, handle and all, about one foot square. This instrument is pointed at a particular moving object, and by turning a handle several photographs are taken each second. These are converted into transparencies and placed in succession upon a strip which is wound on rollers passed through a lantern of peculiar construction, and by its agency projected upon a screen.

Friese-Greene even had ideas about adding sound to moving pictures. Hearing of Thomas Edison's success with recording and playing back sound with his phonograph, the Englishman wrote to the American inventor, suggesting a collaboration.

WHO REALLY INVENTED THE CINEMA?

It didn't happen and, in 1889, Friese-Greene was committed to prison for debt. All his cameras and other apparatus were sold by public auction to defray his debts. On his release from prison, he continued with his experiments, but never had the funds or the contacts to publicise his work in the way it deserved. When he died in 1921, the British film industry paid its tribute to him by financing his funeral and gravestone in Highgate Cemetery.

Wordsworth Donisthorpe
Wordsworth Donisthorpe was the son of an inventor, and a Cambridge University graduate who became a barrister, though he never practised. Instead, in 1876, he took out a patent for a camera that exposed a series of glass plates in rapid succession. The idea was that prints made from the plates would be pasted onto a paper strip and viewed in similarly rapid succession by an electric spark. He called his invention the Kinesigraph, and, like Friese-Greene, even considered linking the moving images with sound, supplied by a phonograph.

In 1889, Donisthorpe took out a new patent on another more practical machine, which this time exposed its series of images on sensitised strips of celluloid. In order to expose each frame of film, a shuttle carrying the film moved upwards as the film was pulled down, so that the film was stationary relative to the lens during the brief time needed for exposure. The Kinesigraph name was also used for this second, more feasible, apparatus. A projector to show the moving images so produced was also mentioned in the patent.

Today, all that remains of Donisthorpe's experiments are ten frames of celluloid film shot, using the Kinesigraph, in Trafalgar Square in 1890, probably the earliest surviving film fragment of London.

Louis Le Prince
The introduction of paper roll film by American photographic pioneer George Eastman in 1885 had an impact on the work of French inventor Louis Le Prince. He proposed the use of flexible paper film moving through a camera by means of an intermittently revolving take-up spool. The main problem with the idea, however, was that, as the take-up spool became fuller and so correspondingly fatter, the exposures made on the paper became more and more separated. But still surviving short lengths of film made in 1888 show that, in theory at least, his idea worked.

THE INGENIOUS VICTORIANS

Thomas Edison

Of all the inventors associated with the birth of cinematography, Thomas Edison was the most renowned, and so it is his name that is most often linked with the first cine camera.

At the time that Edison began his experiments, he had already achieved success with his phonograph, which recorded and played back sounds on a rotating cylinder. So when it came to animating photographic images, it's not surprising that his early attempts involved placing a series of tiny images on a sensitised, intermittently rotating cylinder.

George Eastman (left) and Thomas Edison working together.

He soon switched his attention, however, to a more practical base for his images when he heard about the first roll film snapshot camera announced by George Eastman in 1888, which used celluloid film. Using cut-down lengths of Eastman's still camera film, Edison produced a camera and projector combined, which he called the Kinetograph.

WHO REALLY INVENTED THE CINEMA?

Remarkably, it seems that, when used as a projector, Edison's Kinetograph was linked to a phonograph to provide sound. In 1891, the year Edison applied for a patent on his device, an English magazine called *Photography* reported:

> The Kinetograph is a machine combining electricity with photography, so that a man may sit in his drawing-room and see reproduced upon a screen the stage of a distant theatre, may observe the actions of the actors exactly, and also hear the voices of the players or the music of an opera. So exact is the instrument that every muscle of the face and every expression is faithfully reproduced. The machine will, for instance, reproduce a boxing match. The whole scene is reproduced, every blow struck is seen perfectly and even the sound of the blows can be heard.

Despite this early success, which, according to the press report, projected a moving picture onto a screen, Edison then moved in a slightly different direction.

In 1893, when he eventually demonstrated his apparatus to the public, there was no sound accompaniment and the moving images were viewed, not by projection from the Kinetograph, but in the Kinetoscope, a separate electrically driven device in which the moving picture could be viewed only by one person at a time. His first commercial showing of motion pictures came the next year, in 1894, with the opening of a 'peephole' Kinetoscope parlour on Broadway in New York City.

It was American inventor Thomas Armat who took things a more logical step forward. As Edison moved on to other ideas that had taken his interest, Armat acquired the rights to build and improve on Edison's Kinetopscope to produce his own version, which he called the Vitascope. He used this in 1896 for the first true demonstration of the cinematic art to a mass audience at New York City's Koster and Bial's Music Hall.

Birt Acres

Meanwhile in Britain, American-born Birt Acres was treading a similar path to Edison. In fact, he claimed he had invented something very similar to Edison's apparatus, which he had been using for several years before Edison's announcement, although he had not applied for a patent. In a letter to the British magazine *The Amateur Photographer* in 1897, he stated:

THE INGENIOUS VICTORIANS

> The fact is I feel there is a great future for animated photography, but there is also great room for improvement, and as I have my heart and soul in the subject, it is not to be wondered at that, instead of seeking a bubble reputation as a music hall showman, I have been endeavouring to perfect my system.

In fact, three years before, in 1894, Acres had used a device of his own design, which he called the Kinetic Camera, in an attempt to film the opening of the Manchester Ship Canal. The attempt was unsuccessful, although he did manage to obtain images of that year's Henley Regatta. After that he went on to become what many consider to be the first newsreel photographer.

The Birtac, a camera that introduced cinematography to amateur movie makers.

His other claim to fame was the invention of a camera combined with projector that, according to its handbook, 'placed animated photography within the reach of everyone.' The complete system, launched in 1898, was called the Birtac Home Cinema, and it was the first camera to be aimed at amateur movie makers.

WHO REALLY INVENTED THE CINEMA?

An 1899 advertisement for the Birtac camera.

The Lumière brothers

Unlike still photographers, whose images could be produced directly from the film or glass plates that had been exposed in the camera, cinematography needed another form of apparatus to project the moving images. In short, the history of the cinema was never going to progress without the invention of a practical method of projection. Although most movie camera inventors had considered ways to project their images, often using the same apparatus to shoot and project, the inventors most closely associated with a practical form of film projection were Auguste and Louis Lumière, in France.

During 1894, the year after Edison's announcement, the two brothers worked on their projector project before announcing the hand-cranked Cinématographe. Perhaps the most important improvement on Edison's designs was in the way the film moved through the Lumière apparatus. Edison had run the film continuously. The Lumière brothers saw the

THE INGENIOUS VICTORIANS

The Lumière brothers, pioneers of film projection.

The Cinématographe, invented by the Lumière brothers.

advantage of running it intermittently. In the Cinématographe, the film was advanced through the gate by a pair of claws that caught the film in its sprocket holes and pulled it down one frame at a time. It was based on a sewing machine movement, and so the Lumière brothers became the first to demonstrate the claw system, which subsequently became the standard for cine camera and projector design

Early films
In these early days, film projection was a dangerous business. Early film manufacture involved the use of nitric and sulphuric acid, with the addition of alcohol or camphor. The so-called nitrate film that resulted from this somewhat toxic mixture was dangerously flammable, especially when run though a hot projector.

Even when the film wasn't being projected, its chemistry was such that it could quickly decompose and spontaneously combust. If nitrate film did catch fire, whether from a heat source or spontaneously, it burnt rapidly, fuelled by its own oxygen, whilst releasing toxic fumes, and was not easily extinguishable by water. For this reason the projection booths of early cinemas were likely to be asbestos lined.

Nitrate film was mainly confined to the 35mm gauge used by professional film-makers and cinemas, but as amateur film-making began to grow in popularity, it was also used in early cut-down versions such as 17.5mm. It was a brave man, or woman, who took employment as a professional film projectionist or who took up amateur movie making as a hobby in Victorian times.

Home movies
In these early days of cinematography few Victorians would have thought of owning a camera to shoot movies. But they were interested in seeing the results of professionally shot films.

One way of doing that would have been with a Mutoscope, which used a series of images, shot in sequence by a camera called the Biograph, made by the American Mutoscope and Biograph Company, founded in 1895. This used a mechanism whose rollers wound film past a lens, intermittently stopping for the exposure each time a gap in the rollers' outside edge came around. Prints made from the resulting negatives were attached by one edge to a drum within the Mutoscope and viewed in succession as a handle on the outside case was turned.

THE INGENIOUS VICTORIANS

Coin-operated Mutoscopes, whose animated subjects were, more often than not, of a slightly sensuous and innocently erotic nature, became favourites in amusement arcades where they earned the nickname of What the Butler Saw machines. They survived in some places until as late as the 1970s.

In 1897, shortly after introducing the Cinématographe, the Lumière brothers went on to produce a small home entertainment machine called the Kinora. The device consisted of a handle connected to a wheel with a series of sequentially shot small pictures attached. These were viewed by one person at a time, looking down onto the pictures through a viewing hood. As the wheel was turned, the pictures were flipped one at a time to give the illusion of movement. Each reel of pictures gave about twenty-five seconds of action, according to the speed the handle was turned.

The Kinora home movie machine with its film reel.

The Mutoscope, popularly known as a What the Butler Saw machine.

WHO REALLY INVENTED THE CINEMA?

The design was later remodelled by the English Kinora Company when professionally shot reels were supplied to rent or buy for home viewing. Users could also have motion portraits of themselves shot in a special studio to view on the Kinora. The device remained popular as an amateur form of home movie entertainment for about twenty years.

Although several inventors experimented with adding sound to their moving images, the Victorian cinema remained mostly silent. It wasn't until more than a quarter of a century after Queen Victoria's death that Warner Brothers, in 1927, released the first film with synchronised dialogue. It was called *The Jazz Singer*, in which the star Al Jolson first uttered the now famous words: 'You ain't heard nothin' yet.'

CHAPTER 20

The Age of the Velocipede

The Victorians didn't invent bicycles, but it was in the Victorian era that they acquired pedals and came of age. Except that they didn't call them bicycles at first. To the Victorians this was a mode of transport initially referred to as the velocipede, a word taken from the Latin meaning 'fast foot'. It's also worth noting that the word 'bicycle', by definition, denotes a cycle with two wheels. It wasn't uncommon for a Victorian velocipede to have three or four wheels – and some had only one!

The first vehicles to be powered by human effort date back to the seventeenth century. But the first human-powered two-wheeled land vehicle to gain public acceptance was the brainchild of German inventor Baron Karl von Drais in 1817. His device comprised little more than a wooden frame with a wheel front and back, on which the rider sat and used his legs to scoot himself along the road. He called it his Laufmaschine (running machine), also known as the Draisienne or Mechanical Horse, but the name that caught on with the public was the Dandy Charger.

Pedals on the wheels
By the start of the Victorian era, it began to be obvious that if such vehicles were to evolve, there should be less scooting and more pedalling involved. In 1839, Scottish blacksmith Kirkpatrick Macmillan designed and produced a two-wheeled velocipede that was driven by the rider pumping foot treadles whose push rods were connected to crank pins on the rear axle to rotate the rear wheel.

French mechanic Pierre Lallement was among the first to see the benefits of substituting pedals for treadles, but rather than attaching them to the wheel via a chain, as would be the case in later bicycles, he placed the pedals directly on the front wheel, one that was just slightly larger than the rear wheel.

A selection of the kind of velocipedes that won popularity during the Victorian era.

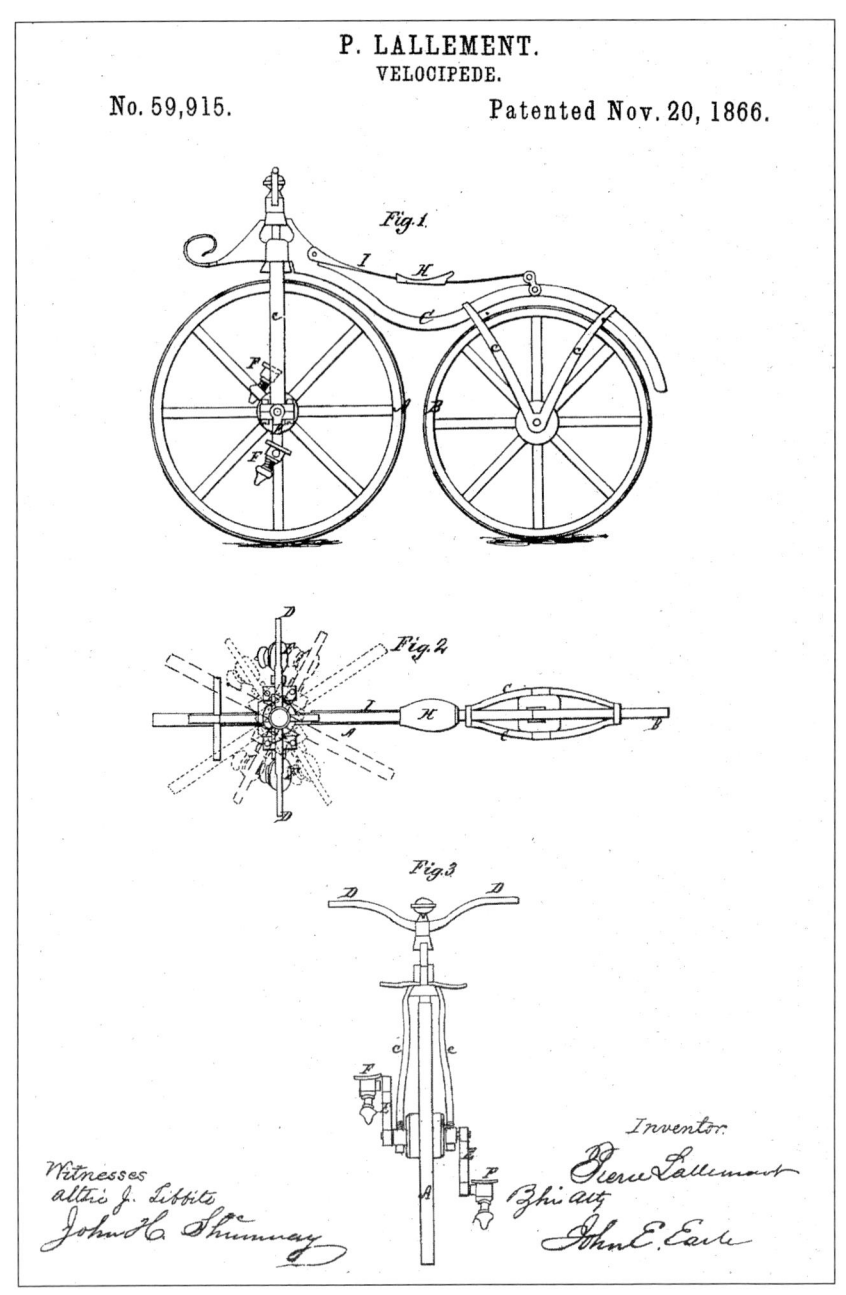

Pierre Lallament's patent of 1866.

THE AGE OF THE VELOCIPEDE

This was a style that continued to be favoured by a great many velocipede inventors and manufacturers over the following few years until the mid-1870s, when the high wheel bicycle that became known as the Penny Farthing became immensely popular, at about the same time that the name 'bicycle' began to take over from 'velocipede' for many manufacturers. The name was based on the old English pre-decimal currency, which contained a penny (of which there twelve to the shilling, with twenty shillings to the pound) and a farthing (a quarter the value of a penny), two coins of very different sizes.

The Penny Farthing featured a front wheel whose diameter was about four times that of the back wheel. It was onto this huge front wheel that the pedals were attached, with a seat and handlebars above. A curved bar ran from the top of the fork that held the front wheel, under the saddle and down to the small back wheel that rotated at its end.

The now almost laughable size of the front wheel of a Penny Farthing was made as large as possible to increase the machine's speed and to reduce road shocks. The downside was that the machine was difficult to mount and devilishly hazardous to ride. If that large front wheel happened to strike an obstacle in the road, the whole apparatus rotated forward and the rider, often with his legs trapped under the handlebars, hit the road head-on and from a great height.

That was a concern addressed, among others, by the American Eagle Bicycle Manufacturing Company. The company's Eagle Bicycle reversed the wheels so that the rider sat above, and peddled, the back wheel, whilst steering with handlebars attached to a much smaller front wheel. The company's advertisements made

How the Eagle Bicycle was advertised.

great play of their machine's safety, strength and speed, illustrating one of their bicycles being safely ridden alongside someone on the more traditional Penny Farthing about to fly over the handlebars and big front wheel. 'Old habits die hard,' said the poster.

Pedals with chains
It wasn't until 1879 that pedals began to be removed from the front or back wheels and placed instead between them, connected to the rear wheels by a chain. British bicycle designer Henry Lawson was among the first with a machine that he called the Bicyclette. Lawson had previously produced a treadle-driven cycle, which he now replaced with rotary pedals attached by a chain to a rear wheel about half the size of the front wheel. By making the sprocket in the rear wheel smaller than that

Riding a safety bicycle didn't necessarily guarantee your safety!

THE AGE OF THE VELOCIPEDE

Two wheels, pedals and a chain to drive the rear one, as shown in Thomas Jeffery's patent of 1889.

on the pedals, he evolved a system in which the rear wheel revolved more than once for every single revolution of the pedals, thus giving more speed.

    As other inventors and designers took up the idea of pedals and chains, this new kind of machine became known as a safety bicycle, largely because, with the rider's feet within easier reach of the ground, rather than the distance in previous designs, it was so much safer to ride.

THE INGENIOUS VICTORIANS

Despite this, the new safety bicycles were slow to catch on with the public. Many thought them to be too complicated, others objected to the extra cost of buying such machines, and some even thought that bringing the rider's position so much lower to the ground was undignified.

Gradually, as the end of the nineteenth century approached, most bicycles began to take on a look close to the accepted design of today, with the same size wheels front and back, pedal and chain-driven, and the rider occupying a position midway between the two wheels. Most followed this basic design, but not all. Side by side with the designers and inventors who steered the two-wheeled velocipede towards the safety bicycle, there were others who took their flights of fancy in rather different directions. What mostly differentiated them from the traditional path was in the number of wheels their machines used.

One, three and four wheels

There were those who envisaged a much faster form of vehicle that used only one wheel. One such was the Flying Yankee Velocipede, patented by Richard C. Hemming in 1869. The design comprised one extremely large wheel, about 5 feet in diameter, inside which the rider sat with his feet in stirrups as he turned a crank with his hands. This was linked by belts to a smaller driving wheel that rested on the inner circumference of the large wheel, using friction to rotate it. Direction was controlled by leaning to the left or right while in motion.

Designs for one-wheel velocipedes proliferated, although there were probably more designs suggested than there were machines actually built. In truth, the idea of riding along the road inside one large wheel was never going be as safe or as successful as riding on two wheels, or better still, three.

The tricycle design was a far more practical proposition, and one taken up rather more enthusiastically by lady riders, who could even buy special

Hemming's Unicycle.

THE AGE OF THE VELOCIPEDE

Victor Rotary Tricycle.

Medically advised corsets as advertised for lady bicycle riders.

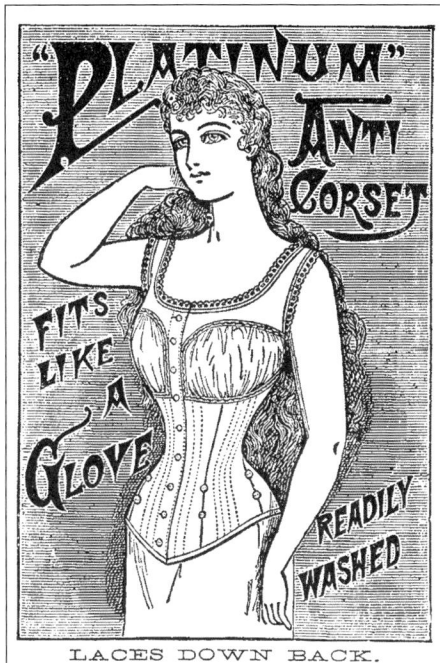

A four-wheeled quadricycle made for two.

corsets designed for women riders. The first three-wheeled velocipedes, with one wheel at the front and two behind, gained popularity in the 1860s. At first, like their bicycle counterparts, the tricycle's pedals were situated on the front wheel. By the 1870s, tricycles began to be made with pedals and chains to drive either the front wheel or the back wheels. Tricycles enjoyed great popularity, and soon two-seaters, known as Sociables, appeared.

The idea of two-seater vehicles became even more popular with the manufacture of quadricycles, which used four wheels. The most obvious place to position four wheels would have been at the corners of the frame, like a cart. But many four-wheelers took on a design that used two large wheels, one each side of the rider or riders, and two more smaller wheels, one at the front and one at the back. The larger wheels were driven by pedals, the front wheel was used for steering and the rear wheel acted as a stabiliser. The riders usually sat one in front of the other, often a man at the back providing the pedalling power, and a lady passenger in front.

Failures and eccentricities
Some developments of velocipede or bicycle were somewhat strange – and some were stranger than others, not just in their design, but equally in the way the vehicles were used.

Monorail vélocipédique.

THE INGENIOUS VICTORIANS

In 1892, American inventor Arthur Hotchkiss patented a kind of upside down bicycle with the wheels on top and the pedals below, that ran along a monorail built on the top of a fence running between the town of Smithville and Mount Holly, just under 2 miles away. Its purpose was to enable factory workers to travel from the town to their factory in about six minutes. For this they paid $2 a month for a ticket. The drawback was that if some cyclists were faster than others, there was no way to overtake, and of course, if two travelling in opposite directions met, they were forced to find a siding for one of them to move into. Nevertheless, Hotchkiss's bicycle railroad operated for about five years before declining interest forced it to close. Similar monorail systems were tried out in England across the counties of Norfolk and Lancashire, but the idea failed to see any great success.

At least two other inventors came up with the idea of running a four-wheeled velocipede along more conventional twin-rail railway tracks.

The Railroad Velocipede, patented by Omer F. Campbell and Frank L. Prindle in 1881, took the form of one large wheel with pedals on it, flanked by two smaller wheels front and back of it running along one track, with a fourth wheel at the end of an axle running along the second track.

In 1896, Charles Teetor patented a rather more conventional design consisting of four wheels of equal

Omer F. Campbell and Frank L. Prindle's patent of 1881.

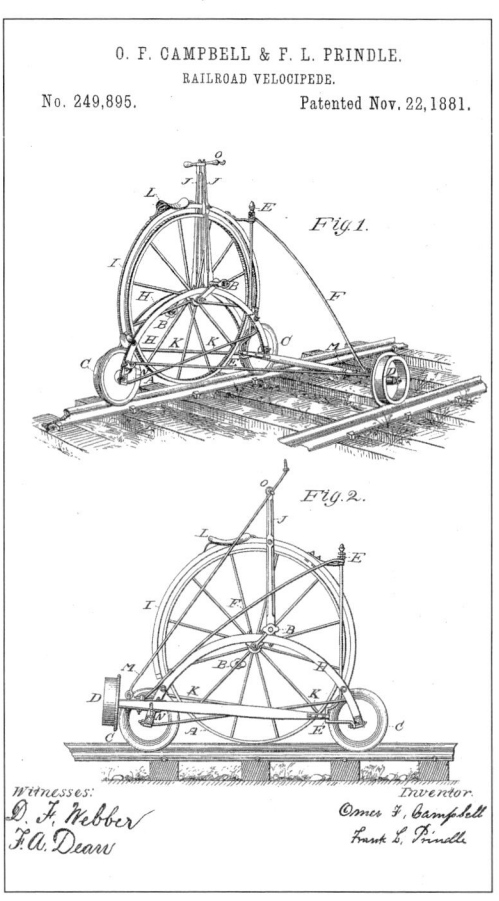

THE AGE OF THE VELOCIPEDE

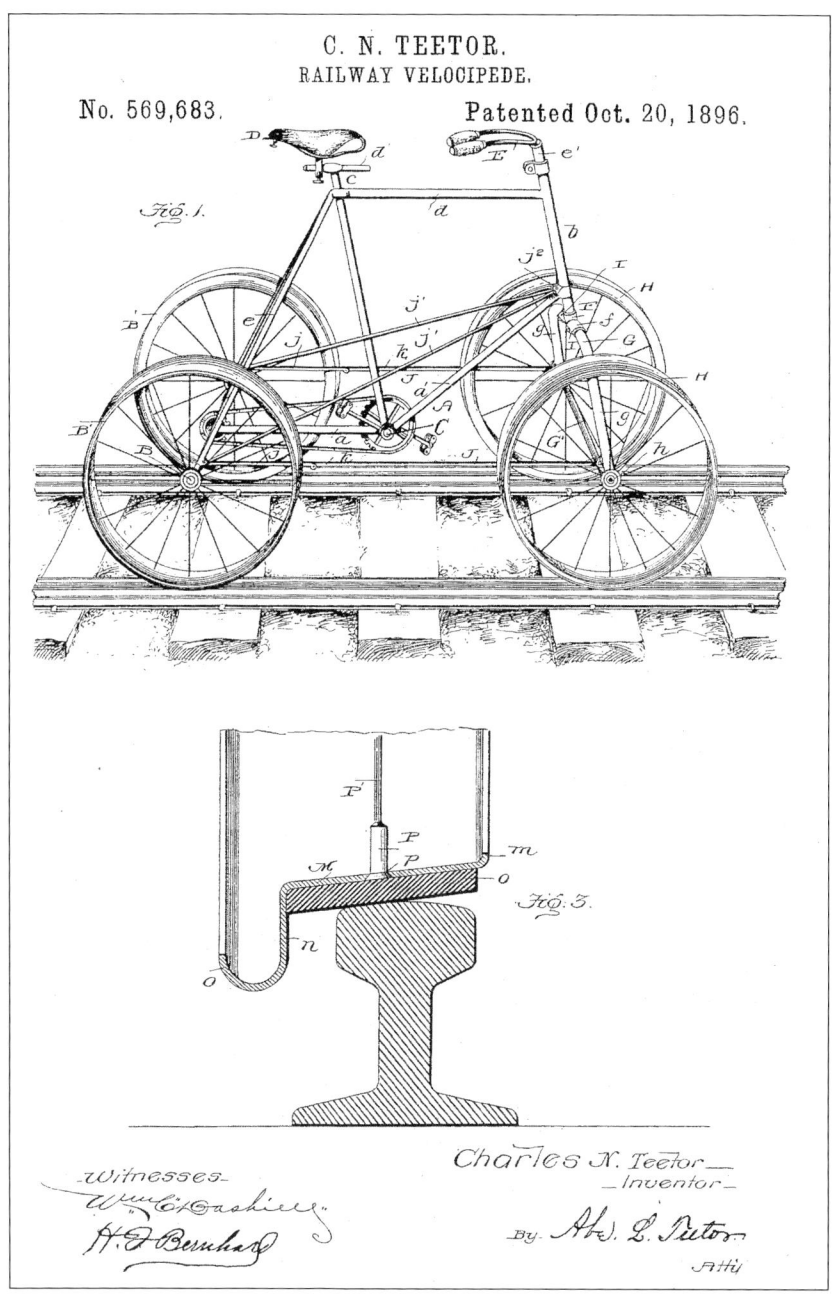

Charles Teetor's patent of 1881.

sizes, two on each side to run on the railway tracks, with a framework between them supporting pedals and a chain drive to the back wheels. A saddle and handlebars were fixed above the frame, although obviously the handlebars were there only for the rider to hang on to, since the velocipede ran on railway tracks, and could not be steered!

The year 1893 saw the brief appearance of the Road and River Cycle, capable of transporting its rider along roads in the normal way, and across water in a rather more unconventional way. This strange contraption took a three-wheel tricycle of the two large wheels at the back and one smaller one at the front design, and fitted it with two small boats between the rider and the large back wheels. These were equipped with paddles attached to the spokes. When the rider reached a river or lake, he simply carried on cycling. The boats kept him afloat and pedalling the paddles on the back wheels propelled him across the water.

Then there were those who saw the coming of the bicycle age as something to be exploited and handled in a completely different way: people like Chas H. Kabrich, who advertised himself as the only Bike-Chute Aeronaut, and who performed a novel and thrilling bicycle parachute act in mid-air.

During the 1880s, a craze began for parachuting from hot air balloons. Unlike those that followed, these balloons did not carry their own source of heat. Instead, the air trapped within the balloon was heated over a bonfire until it took off, carrying the parachutist with it. Kabrich was reputed to have taken this idea one step further, by ascending with, not just a parachute, but with a bicycle attached to it.

If the poster advertising the aeronaut's stunts is to believed, having reached a suitable height, Kabrich then jumped from the balloon, and on the way down, performed various acrobatics on the bicycle.

Nothing more is known of Chas H. Kabrich and it's possible that the only remaining posters printed to advertise his stunts might have heralded his first and last bicycle parachute act.

Chas H. Kabrich poster of 1886.

CHAPTER 21

Pneumatic Railways

At the start of the Victorian era, steam power was behind the rapid development of the railway system. By the end of the era, Britain's first electric railway had been opened in Brighton, pointing the way to the eventual electrification of Britain's rail system, both overground and underground. But between steam and electricity there emerged, for a brief time, another motive force that some saw as the ideal way to power a railway: compressed air. It led to the rise and inevitable fall of the pneumatic railway.

Consider a railway carriage, large enough to accommodate a suitable number of seated passengers. Insert this carriage into an air-tight, tube-like tunnel and use a huge fan to generate enough air to blow it from one end of the tunnel to the other. Then reverse the fan's rotation to produce a vacuum that has the effect of sucking the carriage back again. That, basically, was the theory behind the pneumatic railway.

The idea did not begin as a means of transporting passengers. It was seen initially as a method of moving cargo, and in particular, for transporting Post Office mail. The man behind the idea was civil engineer Thomas Rammell. He proposed his ideas in 1839, which led to a small-scale trial of his system on land near Birmingham, using a type of fan originally designed for coal mine ventilation. The trials were successful and a full-scale experiment was planned using a cast-iron tube made by the Staveley Ironworks in Derbyshire.

The first experiment
The tube was not circular, but took the shape of a traditional railway tunnel, with a height of 2 feet 9 inches and a width that varied between 2 feet 4 inches and 2 feet 6 inches. It was made in 9-foot lengths, each weighing around 1 ton. Raised ledges, 2 inches wide and 1 inch high in

The experimental pneumatic railway test at Battersea.

the base of the tube, were used as rails, on which small, elongated capsule-like carriages ran.

The location for the experiment, which took place in May 1861, was Battersea, in London. The assembled length of tube began at the steamboat pier and ran for 452 yards along the banks of the river Thames, incorporating along the way several twists, turns and gradients. At the far end of the tube, a steam engine drove a 21-foot flywheel that also acted as a fan. When set in motion, it created a vacuum, resulting in the capsule at the Battersea end leaping into life and travelling along the length of the tube in about thirty seconds.

The first experiments were with the capsule loaded with bags of gravel, but the temptation for a person to travel down its length was obviously too great, and soon workmen from the building of the system were squashing themselves into the small capsules to see what the journey was like for passengers. Luckily, it transpired that during the journey, there was enough air in the tube to keep these early passengers alive and relatively well.

Public exhibitions followed and it soon became the vogue for more

Brave passengers prepare for a ride in the freight-carrying carriages that travelled along the tubes of the Pneumatic Despatch Company.

people to risk journeys down the tube. *The Illustrated London News* in August 1861 stated:

> One trip was made in sixty seconds, and a second in fifty-five seconds. Two gentlemen occupied the carriages during the first trip. They lay on their backs on mattresses with horsecloths for covering, and appeared to be perfectly satisfied with their journey.

Following the success of the trials, the London Pneumatic Despatch Company was formed and work began on laying pneumatic railway tubes underground, for use by the Post Office. The first line began at Euston Station and operations to carry mailbags began in February 1863. That same month, *The Times* newspaper reported:

PNEUMATIC RAILWAYS

Between the pneumatic despatch and the subterranean railways, the days ought to be fast approaching when the ponderous goods vans which now fly between station and station shall disappear for ever from the streets of London.

The system carried on transporting mail under the streets of London until 1874, when the Post Office cancelled the contract because the use of the railway was no longer economically viable. The London Pneumatic Despatch Company subsequently closed in 1875.

The Crystal Palace pneumatic railway
Meanwhile, Thomas Rammell had turned his thoughts to a more ambitious project: a full-size pneumatic railway designed to carry passengers in spacious comfort, as opposed to the risky, flat-on-their-backs, claustrophobic journeys hitherto taken by adventurous pneumatic railways passengers.

His chosen site for the first passenger pneumatic railway was the

Entrance to the pneumatic railway that ran for a few brief months at the Crystal Palace in South London.

rebuilt Crystal Palace in South London, where it fitted in well with the many other weird and wonderful attractions of the park.

The railway was built in a 600-yard brick tunnel, 10 feet high and 9 feet wide. The carriage was capable of carrying up to thirty-five passengers, who entered through sliding doors at each end. A fringe of bristles surrounded the carriage and brushed the walls of the tunnel to make the necessary air-tight seal. The air needed to move the carriage came from a steam-driven 22-foot fan at one end.

The carriage began its journey as iron doors closed behind it and air from the fan was blown up through a floor vent to propel the carriage along railway tracks to the opposite end. For return trips, the fan was reversed to set up a vacuum that caused the carriage to move back in the opposite direction, by extracting air through a grill in the tunnel roof.

The Crystal Palace pneumatic railway opened for business in 1864 and was operated for only two months, during which time, passengers were charged 6d (2½p) for each trip.

The Waterloo to Whitehall pneumatic railway
The relatively short life of the Crystal Palace railway was, however, little more than a prelude to a much more ambitious project to build a pneumatic railway from Waterloo, under the river Thames to Whitehall. Carriages were expected to run one every three or four minutes, with fares of 2d (a little under 1p) for first class travellers and half that for second class. By running about fifteen trains an hour, from 7.00 am to midnight, it was calculated that the railway should make something like £23,268 per year, enough to give investors a good return on their money.

The 22-foot diameter fan used to propel the carriage was to be situated at the Waterloo side, and used to both blow the carriage to Whitehall and then to suck it back again. Two sets of iron doors would prevent the air being blown into the stations at each end.

Construction began in October 1865. The tunnel was to be made from an iron tube encased in concrete. On each side, cut and cover tunnels were built for the two termini on the banks of the river, from where the tunnel tube ran into the water, supported on brick piers anchored into the river bed.

Construction continued for twelve months. But then came the great financial panic of 1866, following the collapse of wholesale discount bank Overend, Gurney and Company in London, coupled with the abandonment of the silver standard in Italy. As a result, a banking crisis hit London and funding for the railway dried up. Despite appeals for funds and pleas to the government over the next few years, the project was eventually abandoned in 1869.

That really was the end of any real practical attempt to build a pneumatic railway in Britain, although there were those who continued to put forward ideas that failed to come to fruition.

The South Kensington pneumatic railway
In 1877, a company was formed to construct a pneumatic railway between South Kensington Station, which was served by the Underground District Line, and the Albert Hall, about a quarter of a mile away. The man behind it was, once again, Thomas Rammell.

Describing the proposed railway, *The Illustrated London News* in June 1877, stated:

> No doubt our readers will be familiar with the pneumatic system in which passengers are blown through a tube by a gentle gale of two or three ounces of pressure to the square inch in well-lighted, comfortable carriages free from smoke, dust, heat and smell.

It went on to describe the various failures of the system in the past, but added:

> As far as we can ascertain after careful inquiry and investigation of the system, the failures have had nothing whatever to do with the mechanical principles involved. There is no doubt that Mr Rammell can blow a train loaded with passengers through a tube, say, a mile long, with ease, certainty and economy, while the passengers will travel with a comfort as regards ventilation, quite unknown on the steam worked underground railway.

The plan for Rammell's latest pneumatic railway was to solve the problem of passengers who alighted from the Underground at South Kensington Station having to make their way to nearby attractions of the South Kensington Museum, Albert Hall or Horticultural Gardens. The public were happy to pay a small sum to travel the short distance by omnibus, and so a pneumatic railway to cover the same route seemed like it would be a viable proposition.

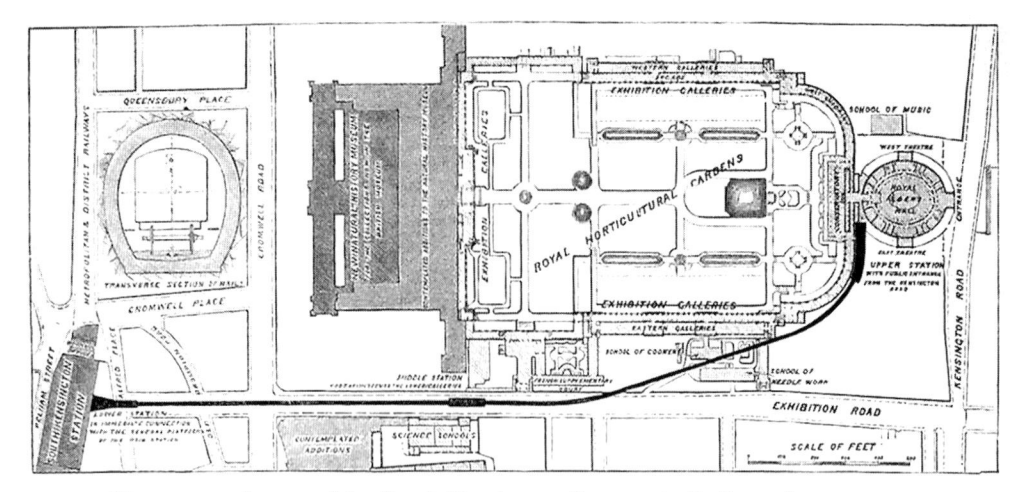

The proposed route of the South Kensington Pneumatic Railway that was never built.

The tunnel was to be made of brick with a paved floor, rising from the underground station to the Albert Hall with a gradient of 1 in 48. The train would be blown along it, not by a fan as had been the case in previous pneumatic railways, but by a huge centrifugal pump driven by two steam engines.

The train would consist of six carriages, travelling on rails with a 4-foot gauge, and capable of carrying 200 passengers. When the train was not running, it was planned that the tunnel would be open to the public to walk its length.

The idea won much support. Rammell got together a board of directors for his company, the Metropolitan and District railway companies supported the plans and a 4½ per cent dividend was guaranteed for investors in the scheme, which required £60,000 capital.

The press were equally enthusiastic. 'No novelty is involved in the scheme and of its mechanical success we have no doubt,' said *The Illustrated London News.*

Despite the drawing of elaborate plans for the railway, and enthusiasm by the press and the engineering fraternity alike, the South Kensington Railway was never built. Thomas Rammell died two years later in 1879, after developing diabetes.

Other pneumatic railways

Another pneumatic railway was planned to travel under the river Mersey, connecting Liverpool and Birkenhead. The tunnel was built, but then plans were changed for it to be used by steam locomotives.

PNEUMATIC RAILWAYS

Tentative unfulfilled plans were even put forward for a pneumatic railway under the English Chanel in 1869. It didn't happen.

The idea was not unique to Britain. Also in 1869, American inventor, publisher and patent lawyer Alfred Beach began building a pneumatic railway under Broadway in New York City. His Beach Pneumatic Transit Company succeeded in building a tunnel 312 feet long in just fifty-eight days. Beach initially put the scheme forward as a system for transporting mail, but by the time it was completed he had opened it to passengers with the idea of extending the line to Central Park.

At the time of its opening, however, there was no real final destination for the railway, and so passengers could only ride through the tunnel to the end of the line and back again, for no reason other than the actual experience. Nevertheless, interest was high at the start, but delays in getting permission to extend the line to anywhere worth visiting, coinciding with a stock market crash that halted any potential investment, meant that public interest eventually waned and the railway was closed in 1873.

Pneumatic railways, on both sides of the Atlantic, had reached the end of the line.

Tunnel entrance to the Beach Pneumatic Railway in America.

CHAPTER 22

The Rise and Fall of Electric Submarines

On the basis that electricity and water are not the best of companions, particularly when people are in close proximity to each at the same time, it would seem that an electrically powered submarine might not be the way forward for undersea exploration. Victorian inventors thought otherwise.

There was, at this time, nothing new about the actual theory of building a submarine. As early as 1580, amateur scientist William Bourne suggested a way of making a boat capable of travelling underwater by use of expanding and contracting chambers to decrease the volume of a vessel to make it sink, and then to increase the volume, causing it to rise again. In 1623, Dutch innovator Cornelis Drebbel built what is usually recognised as the first submarine, which was little more than an enclosed rowing boat powered by twelve oarsmen on a voyage 15 feet below the surface of the river Thames. Other ideas were put forward by numerous English, French, German and American inventors over the following years, some of which were actually built, others existing only as proposals on paper.

Into the Victorian era, there came designs from Prussian Army Colonel Wilhelm Bauer for a sheet iron submarine powered by a two-man treadmill; one from American shoemaker Lodner D. Phillips for a hand-cranked, one-man submarine; another from Wilhelm Bauer for a sixteen-man vessel; one from French engineer Brutus de Villeroi for a vessel 46 feet long, propelled by sixteen oarsmen and a 3-foot diameter hand-cranked propeller … and these examples are only a few of the many. Some of the designs enjoyed success, others sunk without trace.

But then came Claude Goubet with his design for the first electric submarine. Born in Lyon in 1837, Goubet became an inventor who filed

THE RISE AND FALL OF ELECTRIC SUBMARINES

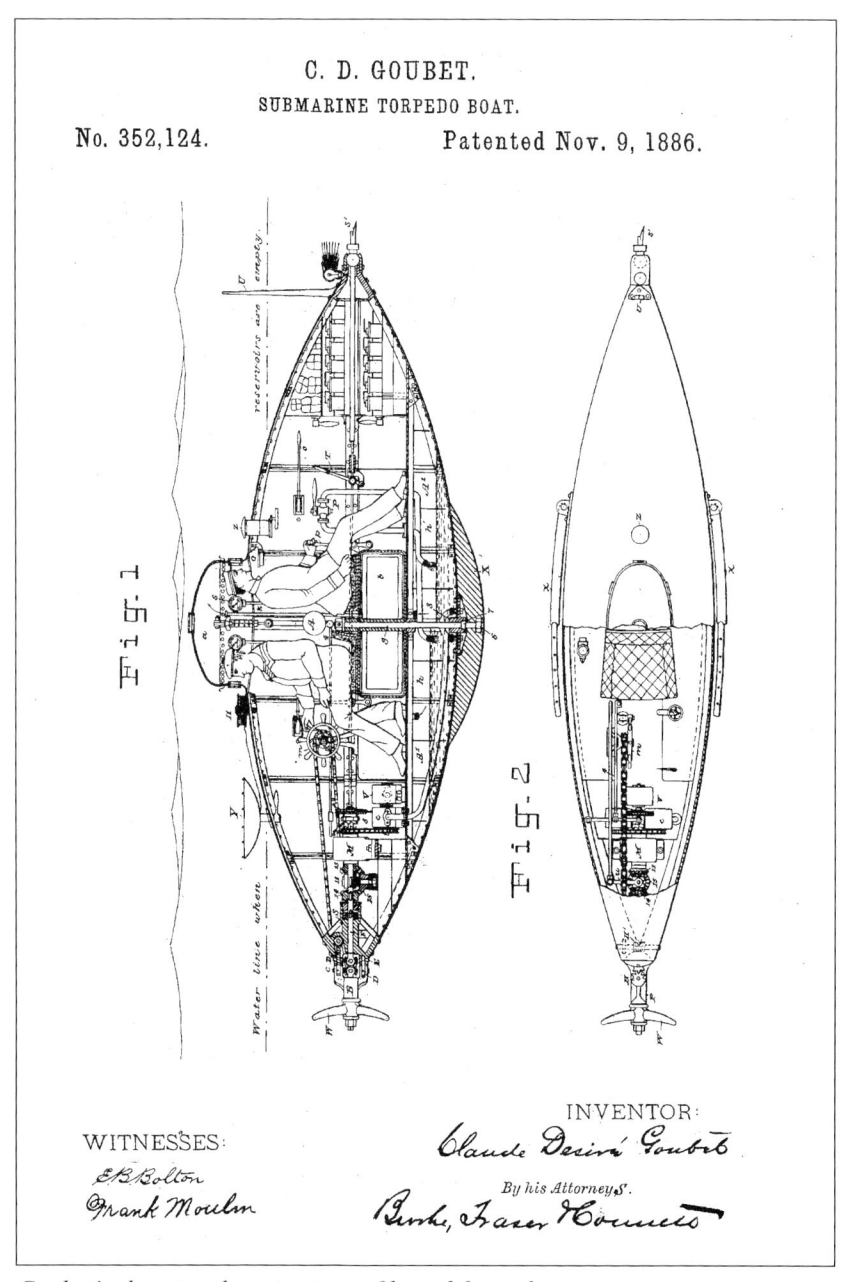

Goubet's electric submarine in profile and from above.

a great many patents in the field of mechanics before getting interested in submarine design in 1880 and beginning to look at actual construction in 1881. Four years later, he filed a patent for his vessel, which he called the *Goubet I*. It was launched in 1887.

The *Goubet I* was cigar shaped with a raised dome in the centre, through which the two-man crew entered and was then hermetically sealed with a cap. The two men sat on seats, back to back, with their heads in the dome. Small glass peepholes that allowed them to see out could be quickly sealed with caps if the glass was accidentally broken.

In the base of the vessel there were watertight chambers. When water was allowed into either or both of these chambers, the submarine sank. When a motor was used to pump the water out again, it rose to the surface.

Above the tanks and below the seats on which the sailors sat, there was a chamber full of compressed air. Pipes from the compressed air chamber led to the top of the dome, where they discharged air to the crew close to their faces. Stale air was evacuated by an air pump.

A motor, connected to a drive shaft, was used to drive the propeller and so move the vessel through the water. Alternatively, one of the two-man crew, sitting in extremely cramped conditions, was able to turn a crank to rotate the propeller. The propeller could be angled left and right to steer the vessel without the use of a rudder. A third method of propulsion came from oars that passed from the inside to the outside through watertight apertures.

To keep the submarine level in the water, a swinging pendulum was connected to pumps that drove water into tanks at the front or back of the vessel, depending on its angle in the water. If the submarine tilted backwards or forwards, the pendulum compensated to activate the appropriate pump, the weight of water pumped into the tanks quickly stabilising the angle of the vessel.

The motor that drove the propeller, the water pumps and air pumps was electric; its power came from accumulators stored in lockers at the front of the vessel.

The *Goubet I* was also equipped with an electrically exploding torpedo, so that the submarine might be used to attack enemy ships. According to *Goubet's* patent, the torpedo was attached to the back of the submarine's dome by means of a fastening that could be operated from the inside. A long wire ran from the torpedo and onto a drum.

For the purposes of attack, one of the crew sighted and lined the submarine up with the enemy ship before submerging, noting the bearing on a compass. The vessel was then submerged, and using the compass bearing, headed for the target. Once it reached the enemy ship, a crew member caused the submarine to submerge to a depth that took it under the ship. Once in place, the torpedo was released to rise in the water and fix itself to the underside of the hull above.

The submarine then moved away, paying out the wire from the drum to the torpedo as it went. The number of revolutions of the drum indicated to the submarine crew the distance that they had moved. When the submarine had positioned itself at a safe enough distance, one crew member closed an electric circuit inside the dome, which sent an electrical charge along the wire to fire the torpedo.

Initially, the French naval and military authorities took a keen interest as the *Goubet I* was tested and demonstrated, first in the river Seine in Paris and later in Cherbourg and Toulon harbours. In the end, however, Goubet's submarine was rejected as being too small and he was invited to submit a design for something lager.

The *Goubet II*, powered by a more powerful electric motor, added three fins to improve stability and had provision for an extra torpedo. Despite the improvements and with problems from backers, the *Goubet*

Launching the Goubet I.

II suffered the same fate as the *Goubet I*. The inventor was ruined, and died in 1903.

When the French authorities rejected the Goubet submarines, they turned their attention to another electrically powered submarine called the *Gymnote*, launched a year after the *Goubet I*, in 1888.

The *Gymnote* was based on experiments by French naval architect Stanislas Charles Henri Dupuy de Lôme, who died in 1885, after which his work was carried on by marine engineer Gustave Zédé and automotive engineering pioneer Arthur Krebs. Together, the latter two engineers designed a submarine that was more practical than Goubet's vessels. The *Gymnote* was fitted with the first naval periscope and the first naval electrical gyrocompass. For these reasons, it is the *Gymnote*, rather than the *Goubet I* or *II*, that is more often recognised as history's first successful submarine.

The Gymnote.

Arthur Krebbs designed the electric motor that powered the submarine. The motor was connected directly to the vessel's propeller, which was capable of rotating at 250 revolutions per minute. The batteries to power the motor were in the bow of the submarine, arranged in six banks of cells. In this way the motor developed a maximum of 55 horsepower at 200 volts. The speed with which the motor turned and therefore the submarine travelled was controlled by the number of batteries used and

THE RISE AND FALL OF ELECTRIC SUBMARINES

the ways they were connected to vary the voltage, either in series (positive pole on one battery to negative pole on the next) or parallel (negative to negative and positive to positive).

The hull of the *Gymnote* was made of steel, supported by thirty-one circular frames. Three ballast tanks in the front, centre and rear were filled with water or emptied by electric or compressed air pumps, in order to control the sinking and rising of the vessel.

Although it was the first to use a periscope, the device was soon abandoned due to difficulties with raising and lowering it and inefficient water seals that led to serious flooding on at least one occasion.

Nevertheless, the *Gymnote* made about 2,000 dives during nearly twenty years of service in the French Navy, until 1907, when it ran aground and was seriously damaged. Abandoned in dry dock, a valve was accidentally left open and the *Gymnote* flooded. Repairs were considered to be impractical and it was eventually sold for scrap in 1911.

Half a century later, in France in the 1960s, an experimental submarine was trialled to carry and fire ballistic missiles. It was named the *Gymnote* in honour of what has come to be remembered as the world's first all-electric submarine. Few who award the *Gymnote* this honour remembered the *Goubet I*, a lot less successful than the *Gymnote*, but actually more entitled to the accolade.

CHAPTER 23

Building Big Ben

It should be acknowledged right at the start that Big Ben is not, as many believe, the name of the clock tower that stands beside the Houses of Parliament next to Westminster Bridge in London. Big Ben is, in fact, the nickname of the biggest bell inside the tower, the one that strikes the hours. When it was built, Victorian journalists referred to the tower as St Stephen's Tower in connection with the fact that Members of Parliament in the House of Commons sat in St Stephen's Hall. Officially, the correct name was simply The Clock Tower. But Big Ben is the name that has been inextricably linked with the clock and its tower over the years, so that's the way it will be referred to here.

Big Ben might never have been built had it not been for a disastrous fire in 1834. It started because of the need to dispose of two cartloads of tally sticks, part of an ancient accounting system no longer in use. The unwise decision was made to burn them in stoves situated in a basement under the House of Lords. As a result, in the early evening of 16 October that year, fire erupted and ignited panelling in the House of Lords before spreading through the House of Commons, and with the exception of a few buildings, decimating the Palace of Westminster. Parliament had nowhere to sit.

A competition was launched to design a new building for the seat of the British Government, and ninety-seven entries were received. The winner was number sixty-four, submitted by Charles Barry, an eminent English architect of the time who, coincidentally, had been born in 1795 in Bridge Street, Westminster, right opposite the site of what would become Big Ben. Barry was knighted by Queen Victoria in 1852, the year that saw the completion of the main interiors of the Palace of Westminster.

His initial design for the Palace of Westminster did not include a clock tower, and when it was suggested that one should be incorporated at one end

Westminster Bridge, with Big Ben in the background, as it was in Victorian times.

of the building, it threw him into a quandary, which, in the following few years, led to a number of complications and a fair amount of controversy.

Building the clock

Since Barry was the first to admit that he was an architect, rather than a clockmaker, he needed someone to design the clock that would sit at the top of the tower. His first mistake was to ask the Queen's clockmaker, Benjamin Lewis Vulliamy, for a suitable design, which upset clockmakers across the country who thought that the job should have been put out to competitive tender.

With that in mind, Astronomer Royal George Airy was called upon to draw up a specification – which just made matters worse for the clockmakers who wanted to be considered for the job. Prime among the problems faced by the makers was Airy's insistence that the first stroke of the hour bell should register a time that would be accurate to within one second per day. Given that the clock's hour hand was to be about 9 feet long with a minute hand more like 14 feet in length, and considering both would be open to wind, rain and other atmospheric elements, the aspiring clockmakers unanimously decided that such accuracy was impossible.

Winding the clock.

At this point, barrister Edmund Beckett Denison came into the story. He was appointed by Parliament to act alongside Airy as a referee among the competition entrants. Rather than simply referee the clockmakers, however, he took over the design himself. Despite the fact that he was a barrister, whose knowledge was mainly concerned with the law, he was also an amateur horologist who claimed to be an expert on clocks, locks, bells and architecture. So, although being no more than an amateur in clockmaking, compared to the professional clockmakers who were chasing the contract, Denison took it upon himself to go ahead and design the clock.

It was 1851 before he came up with a design that would satisfy the required specification, and 1854 before the clock was actually completed. It was built by E.J. Dent & Co, a clockmaker founded by Edward J. Dent in 1814, and already famous for the accuracy of the chronometers it manufactured for the Royal Navy as an aid to navigation.

BUILDING BIG BEN

Casting the bell

Next came the casting of the biggest bell – and more controversy. Barry's design had been for a 14-ton bell to strike the hours, but he had not specified how or where it would be built. Once again, Denison interceded and used his somewhat dubious knowledge of bells to instruct the foundry on its specification. He decreed the shape of the bell and the type of metal from which it should be made, both of which were different from the traditional methods and designs of bell making at that time. The work was undertaken by bell makers John Warner & Sons in Stockton-on-Tees. Two furnaces, each capable of melting 10 tons of metal, were specially built for the job. When completed in 1856, the bell weighed 16 tons, but after being transported to Westminster, it unfortunately cracked while under test and had to be broken up.

Undeterred by the fact that his design and composition of metal for the bell might not have been up to the required standard, Denison turned to the Whitechapel Bell Foundry to have it remade. Using metal from the melted down first bell, the foundry cast a new one, taking twenty days for the metal to solidify and cool. At 7 feet high, it was – and remained – the largest bell ever cast by the foundry. After testing, Denison approved the bell and it was transported from Whitechapel in the East End of London to Westminster on a trolley pulled by sixteen horses as crowds lined the streets to watch.

Building the tower

Meanwhile, building of the tower to hold the clock mechanism, Big Ben and all the other smaller bells that made up the chimes had been progressing. It was designed for Barry by his colleague Augustus Pugin, the English architect, designer, artist and critic who was also responsible for much of the interior design of the Palace of Westminster. It was Pugin's last design before he was overtaken by madness and died in 1852.

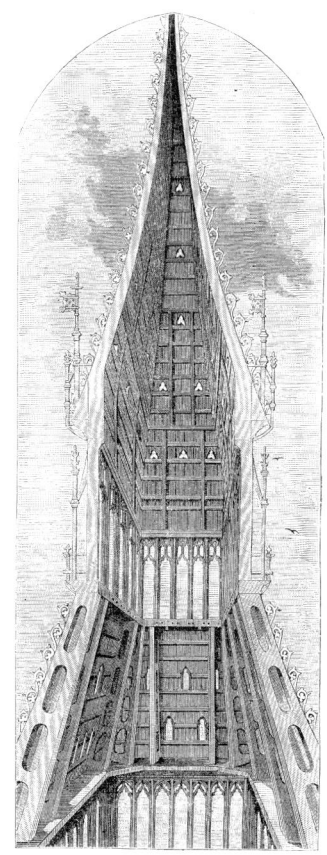

Sectional view of the clock tower.

THE INGENIOUS VICTORIANS

The tower was built from the inside out, so that no scaffolding was seen during its construction. Materials for the job came from around the country: cast-iron girders from London's Regent's Canal Ironworks, stone from Yorkshire, granite from Cornwall, iron roofing plates from Birmingham.

The clock faces were made of opal glass panels set in a 23-foot diameter iron framework with gilded surrounds to the dials. At the base of each clock face, gilt letters proclaimed in Latin: *Domine salvam fac reginam nostram victoriam primam* (O Lord, keep safe our Queen Victoria the First).

Positioning the bell

Falling five years behind schedule, the tower was finally completed in 1859, but not before the task of getting the bells – and in particular, Big Ben itself – to the top of the tower. That in itself took two days.

The bell being manoeuvred into the base of the tower.

BUILDING BIG BEN

On its arrival at the location, the bell was turned on its side and then pushed on a cradle, along rails, with men using levers to manoeuvre it until it came to rest at the bottom of a shaft that ran up the centre of the tower. The 180-foot high shaft was lit by gas and its dimensions were only slightly larger than the bell itself. The cradle was slowly hauled up the tower using a specially made chain, wound by a winch at the top. The winch was hand-cranked, turned by eight men working two handles and took ten revolutions of the handles to wind 1 foot of chain.

The cradle on which the bell sat had four wheels, two at the top and two at the bottom, which ran against guide timbers inside the tower to keep the ascent as smooth as possible.

The bell was hoisted first to the clock room, 190 feet above the ground, where it was turned to its upright position before being finally hauled a further 40 feet up to the bell chamber. The smaller bells, used to ring the quarter hours, had been similarly hauled up the previous week.

There are several theories as to how Big Ben acquired its name. A favourite is that it was named after Sir Benjamin Hall, Chief Commissioner of Works, and a particularly large member of the government who was affectionately known as Big Ben. Another is that the name was derived

Big Ben is hauled up the inside of the tower.

Big Ben and its smaller cousins in place in The Clock Tower.

from Benjamin Caunt, a bare-knuckle heavyweight boxing champion who weighed in at 17 stone and was nicknamed Big Ben. Before the name was adopted, the bell was also known as the Royal Victoria.

On 31 May 1859, twenty-five years after the fire that had sparked the building of Big Ben, its chimes rang out for the first time across London. Then, in September that year, the bell cracked. The problem was due mainly to the fact that Denison, who still claimed to know so much about bell making, had insisted on using a hammer more than twice the weight of that recommended by George Mears, owner of the Whitechapel Bell Foundry.

As a result, Big Ben was taken out of service, and for the next three years, hours were struck on one of the smaller bells. Eventually, the hammer was replaced with a lighter one, the crack was patched with a square piece of metal and the bell was turned slightly so that the hammer fell in a different spot. In that form, it survives today, although even now ringing a slightly imperfect tone, due to the crack.

In 2012, to celebrate the diamond jubilee of Queen Elizabeth II, the clock tower was officially renamed the Elizabeth Tower. Even so, its unofficial name still remains as it has been since Victorian times: Big Ben.

BUILDING BIG BEN

The famous clock face, little different today than it was in Victorian times.

CHAPTER 24

The Amazing Clocks of Charles Shepherd

As visitors flocked to see the many marvels of The Great Exhibition in 1851, it might have been easy to overlook the very first wonder to be seen above the main entrance. It was an unconventional clock whose peculiar shape resulted from its need to match, and incorporate, the architectural design of the Crystal Palace, in which the exhibition was housed.

At its highest point over the main entrance, the Crystal Palace featured a huge semi-circular dome of glass into which the clock was set. Since a traditional round-faced clock would not have been aesthetically pleasing in this position, it was decided to install one with a semi-circular face, which in turn required the use of four hands. It was also unusual for its day for working by electricity.

The Crystal Palace Clock
The hours of the Crystal Palace clock were marked by Roman numerals, starting with VI (six) in the lower left-hand corner of the semi-circular clock face. The numbers then ran in the usual way to XII (twelve) at the top of the clock, then on around the right side of the face to finish with a second number VI in the lower right-hand corner.

The hour hand consisted of two sections pivoted in the middle at the centre of the base of the clock. So when one part of the hand was pointing to XII, the second section was suspended directly below the semi-circular face and overhanging the base of the Palace's glass dome. As the hour hand moved around to I, II, III etc., so the lower part of the hand turned with it. When the top part of the hand pointed to VI on the lower right of the dial, so the bottom part of the hand reached the other VI on the lower left of the dial. This was the hand that then traced the hours up through

THE AMAZING CLOCKS OF CHARLES SHEPHERD

Hands and face of the semi-circular clock designed to stand above the entrance to the Crystal Palace.

VII, VIII, etc., until it reached XII. A similar method was used to indicate the minutes by a second two-part minute hand.

The dial was 24 feet wide, and the hands were made of gilt copper. They were purposely made a little shorter than those of a more conventional clock to prevent their lower parts obscuring a fanlight in the building below the clock as the upper parts pointed to the numbers. The shape of the clock and the positioning of its numbers were echoed in the rest of the dome with thirteen ribs, each one containing one of the clock's thirteen hour numbers, stretching past the perimeter of the clock face and

Shepherd's clock in place above the south entrance to the Crystal Palace.

all the way to the edge of the roof of the dome. The figures were made of zinc, painted white on a blue background, which harmonised with the rest of the building.

The Crystal Palace clock was built by engineer Charles Shepherd, born in 1830 and the son of a clockmaker. Two years before the opening of The Great Exhibition, he had patented a system to control slave clocks from a master clock by use of electricity, also referred to as galvanism, a term used in biology when a muscle is stimulated by an electric current. Along with the unusual semi-circular clock standing above the entrance, he was also commissioned to install more conventional public clocks within The Great Exhibition.

Shepherd was by no means the first to produce an electric clock. In London they were already to be seen at this time in such places as the premises of the Electric Telegraph Company in The Strand and on the front of the Polytechnic Institute in Regent Street. A few were also installed in the private houses of the rich. Mostly, these were run by the use of earth batteries, which comprised twin electrodes or plates of different metals such as iron and copper, buried in the earth, which

THE AMAZING CLOCKS OF CHARLES SHEPHERD

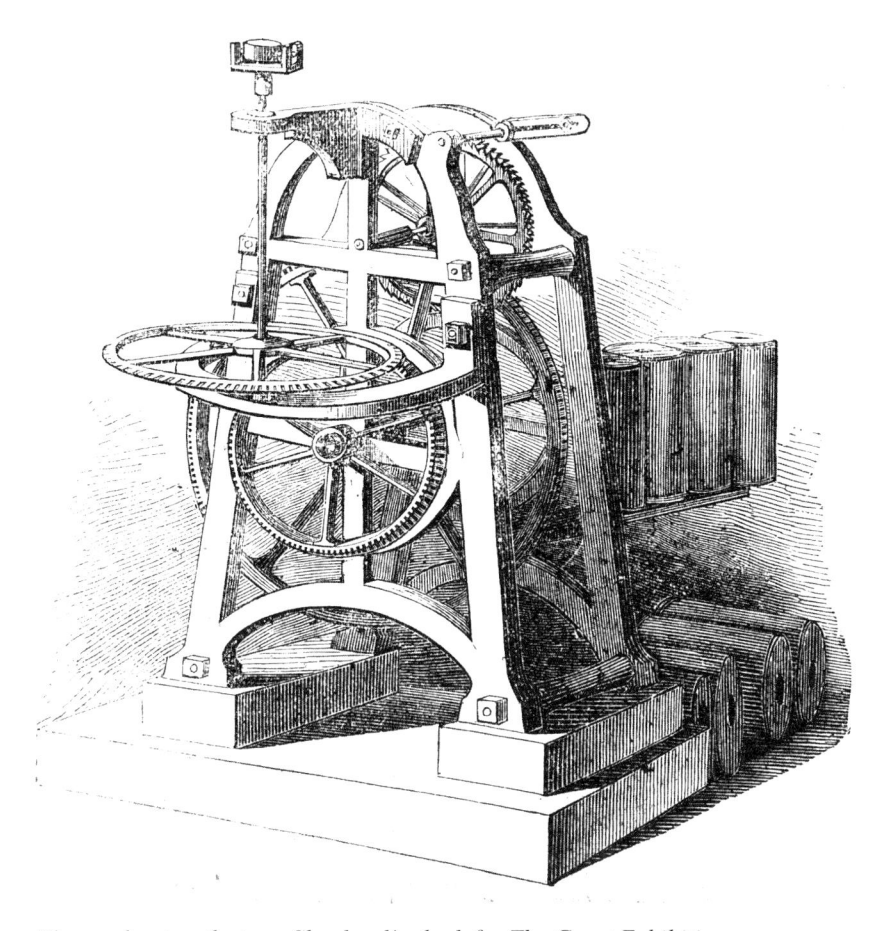

The mechanism that ran Shepherd's clock for The Great Exhibition.

produced a weak electric current. The uncertainty and variances of the amount of power provided by such batteries did not guarantee good timekeeping in clocks powered in this way.

Shepherd improved on this idea by running experimental clocks powered by Smee batteries. Named after their inventor, surgeon Alfred Smee, these were made up of six cells with positive plates made of zinc and negative plates made of silver, immersed in weak sulphuric acid. In this way Shepherd designed electric clocks that ran accurately for up to two years.

It was this type of battery that Shepherd used in his clock for the Crystal Palace. The mechanism that drove the hands was not behind the clock face, as is more often the case. Instead it was placed 48 feet below, with the hands driven from the mechanism by means of several lengths of brass tubing screwed together.

Two other clocks, each 5 feet in diameter, were also installed in the west and east ends of the Crystal Palace and all three were interconnected to keep precise time with one another, governed by a single pendulum that worked by a revolutionary new electro-magnetic system.

The Royal Observatory clocks
The Crystal Palace clock by no means represented the pinnacle of Shepherd's clockmaking career. His next project was a network of interconnected clocks, all controlled by the same master mechanism, for installation at the Royal Observatory in Greenwich.

One of the Observatory's claims to fame is for being the home of Greenwich Mean Time, the place where each new day, year and millennium officially starts at the stroke of midnight. It was Shepherd's clock that helped to establish that, but its genesis lay in the growing network of railways spreading across Victorian England.

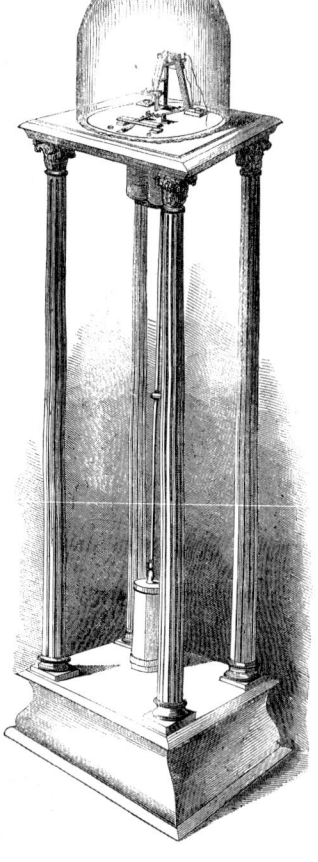

The pendulum that regulated the clocks.

At this time, it was rare for many people to own a clock or a watch, and most relied on public sundials to tell the time. That led to different local times across Britain, with sundials in the east of the country being anything up to thirty minutes ahead of others in the west. But if railways were to run effectively, they needed timetables, and timetables needed a single standard time to be used throughout the country.

Astronomer Royal George Airy was the man who suggested that this standard time should originate and emanate from the Royal Observatory. His idea was that a master clock would be based in Greenwich, from

THE AMAZING CLOCKS OF CHARLES SHEPHERD

Today's copy of Shepherd's Gate Clock at the Royal Observatory.

where it would transmit signals via electric telegraph to slave clocks right across Britain. Charles Shepherd, who had already patented such an idea and shown its success at the Crystal Palace, was the man chosen for the job.

Shepherd's first task was to build one automatic clock to be housed within the Observatory, and from which signals would be transmitted to a large clock face at the Observatory entrance, where it could be seen by the public. He was also commissioned to build three smaller clocks, to be installed inside the Observatory, which would be interconnected with the master mechanism. The Gate Clock featured the usual two hands, but featured a twenty-four-hour face, with hours designated by Roman numerals from 0 at the twelve o'clock position to XXIII at the usual eleven o'clock position.

Signals from the master clock were also to be sent to the time ball on the Observatory's roof. Time balls, which were similarly situated on buildings around the country, consisted of metal or wooden balls that were dropped down the length of a pole at predetermined and accurate times. Often situated near the coast, they were originally used by offshore sailors to verify the accuracy of their chronometers, which were used to determine longitude during their journeys. Shepherd's system included electrical operation of the Greenwich time ball, synchronised with the master clock inside the Observatory, so that its descent ended precisely at 1.00 pm every day.

Shepherd's master clock, later known as the Mean Solar Standard Clock, sent pulses every second to slave clocks at the gatehouse of the Observatory, and to the others inside the Observatory. At the same time, the signals were transmitted by cable to the London Bridge terminus of the South-Eastern Railway, from where signals were sent along all the main railway routes that ran out from London, to a network of slave clocks throughout England. As well as maintaining the accuracy of the time ball on the Royal Observatory, it also sent its signals to another time ball on the office of the Eastern Telegraph Company in The Strand.

The time ball standing on the roof of the Royal Observatory.

In this way, accurate timing was maintained throughout the country and, for the first time, railways were given the means to run on time. Greenwich Mean Time was legally adopted throughout Britain in 1880.

The first electric striking clock

Charles Shepherd was also responsible for the country's first electric clock with a striking mechanism. It was installed in the tower of the Church of St Peter at Hornblotton in Somerset, in 1883. The clock was connected by an electric cable to a pendulum in the basement of nearby Hornblotton House. The clock mechanism included two large coils that formed electric motors that activated levers to pull cables and so ring bells on the hour and on each quarter.

The estimate of cost supplied by Shepherd in April 1883 spoke of a compensating pendulum fitted with a magnetic apparatus for correcting

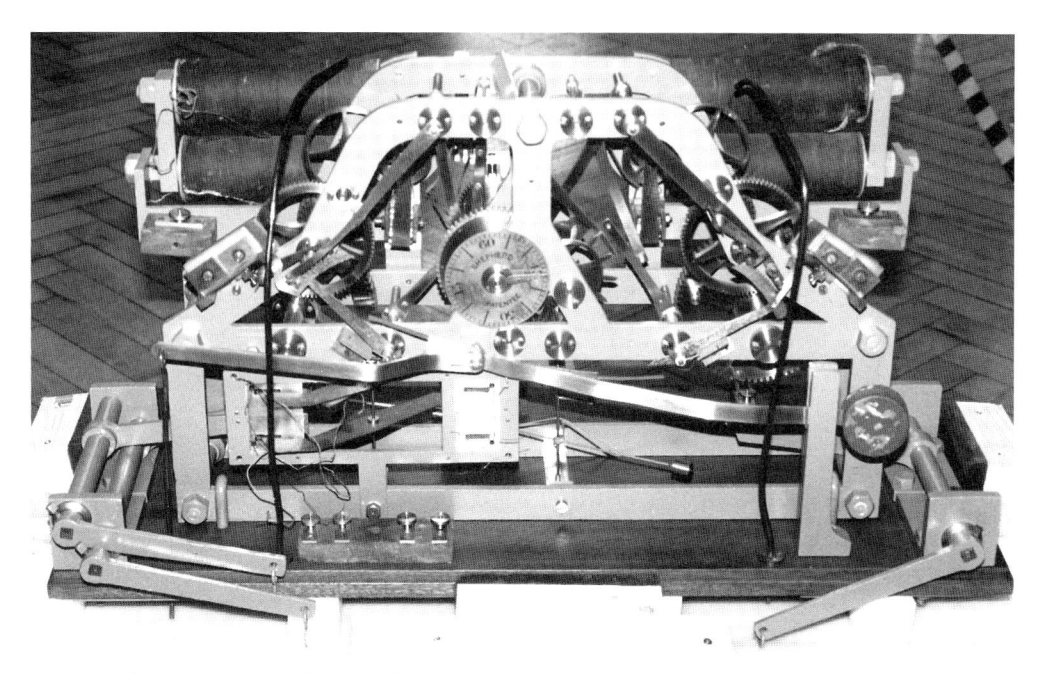

The surviving mechanism from Shepherd's first electric striking clock.

small errors of time, to be housed in a glass-fronted mahogany case. The mechanism that turned the hands was also to be housed in a mahogany case. The clock dial was 2 feet 3 inches in diameter and was to be made of painted copper or iron.

The hours were struck by a bell 2 feet 6 inches in diameter, whilst the quarters were struck by two bells 2 feet 3 inches and 2 feet 1 inch in diameter. The hammers that struck them were powered by electric wires from the clock. The batteries supplied with the mechanism were reckoned to last for four years.

Shepherd's quote for building, supplying and installing the clock was £193.

Today, little remains of Shepherd's pioneering work in clockmaking. When the original Crystal Palace in Hyde Park was dismantled and rebuilt in South London, Shepherd's clock did not go with it. His Gate Clock at the Royal Observatory was destroyed by a bomb during the Second World War and the copy that stands in its place today is controlled by a quartz mechanism. The pendulum for the striking clock at Hornblotton no longer exists, but the rescued mechanism from the clock tower is in storage at the Science Museum in London.

CHAPTER 25

From Today, Painting is Dead

Two years into the Victorian era, the first viable method of photography became a reality, along with the first commercially available camera. It involved a complicated process that was, at first, adopted purely by professional or the most enthusiastic of amateur photographers. By the end of Victoria's reign, the general public, most of whom had no practical expertise in photography, were buying and using the first snapshot cameras. All this led to a fascination among the Victorians for having their portraits taken, and the number of different photographic processes that came and went during this time meant that the resulting photographs took many different forms.

What has come to be acknowledged as the first photograph was produced in 1826, by Joseph Nicéphore Niépce, who successfully recorded the image of a French courtyard. His process involved coating a pewter plate with bitumen of Judea and then placing it in a specially made camera, adapted from a camera obscura, an implement with a lens that projected its image onto a screen that artists used as an aid to composition and perspective. The exposure took all day. Niépce is to be applauded for his achievement, but his process was totally impractical if photography was to move forward.

Daguerreotypes
The man who succeeded in making photography – and portraiture – viable for the first time was another Frenchman, Louis Jacques Mandé Daguerre.

Daguerre was a talented artist and an apprentice architect who became a scenery painter, then scenery designer for the Paris Opera. He was also an acrobat, tightrope walker and an exponent of the diorama, a popular theatrical experience that involved painting scenes on translucent screens to create three-dimensional illusions by the use of clever lighting.

As an artist, Daguerre used a camera obscura, and like Niépce before

FROM TODAY, PAINTING IS DEAD

him, became intrigued with trying to capture and retain the image that the device projected onto its screen. In 1839, he announced his daguerreotype process to the French Academy of Sciences. Legend has it that French artist Paul Delaroche, on seeing his first daguerreotype, exclaimed: 'From today, painting is dead.'

The camera that used the process was designed by Daguerre. It was made of two boxes, one of which slid inside the other. An image was focused on a screen on the back of the camera by sliding the inner box, containing the focusing screen, backwards and forwards inside the outer box, which held the lens. Once focused, the metal plate on which the image was to be recorded replaced the focusing screen.

The daguerreotype process, by which the images were made, was long, complicated and somewhat dangerous. It involved a plate of copper coated with silver, which by the light of a candle and in an airtight box, was exposed first to iodine crystals, then to bromine fumes, before being placed in the camera and the exposure made by removing and replacing a flap across the lens. For development, the exposed plate was placed back in the airtight box as fumes from mercury, heated in a dish over a lamp, reacted with the sensitised plate to form an image.

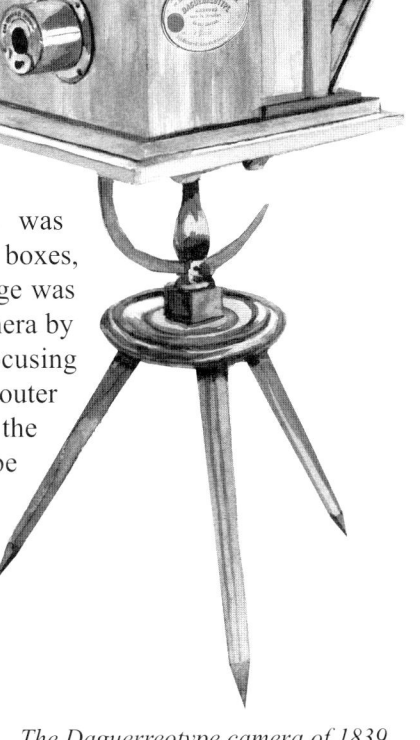

The Daguerreotype camera of 1839 introduced the first truly viable method of photography.

The result, once it had been washed with a salt solution, was an image on the silver-plated copper in which light areas were white and dark areas were represented by the highly reflective silver. Depending on how light struck a daguerreotype, it might have been seen as a positive, a negative or even a combination of both.

Exposure times at first were extensive, but with improvements made to lenses, equipment and the chemistry, it wasn't long before exposures

Daguerreotypes were the first real photographs.

were shortened enough for photographic portraits to be taken for the first time. As many more camera makers began manufacturing cameras for the daguerreotype process and more photographers became adept at handling it, the first portrait studios opened for business.

Even so, having a portrait taken with the daguerreotype process was not an easy procedure. Exposure times might have been shortened considerably, but they still required the sitter to stay very still, sometimes for minutes. For this purpose, special chairs were made with clamps that held the sitter's head in position to prevent movement.

Calotypes

Meanwhile in Britain, William Henry Fox Talbot, squire of the manor in the English village of Lacock and amateur scientist, had been carrying out his own experiments using sensitised paper. While Niépce and Daguerre had taken what must have seemed the logical step of producing a direct positive image, Fox

A chair with head clamps used to keep subjects still when daguerreotype portraits were taken.

246

FROM TODAY, PAINTING IS DEAD

A calotype portrait by pioneering British photographer Roger Fenton.

Talbot recognised the importance of producing an intermediary negative, from which many positive prints could be made.

Initially, Fox Talbot's pictures were made by a process called photogenic drawing, in which the image was exposed until it began to actually appear on the sensitive paper, necessitating very long exposure times, totally unsuitable for portraiture. Even so, that was a problem that he approached in a new way. Rather than build larger lenses in an effort to reduce exposure times, he used smaller cameras, in which the lens's light was concentrated on a smaller area and was therefore brighter. The tiny cameras were built for him by a local carpenter in Lacock and equipped with microscope lenses. Fox Talbot's wife Constance gave these miniature models the name of mousetrap cameras. In 1835, four years before Daguerre's announcement, Fox Talbot used one to produce the world's first negative.

Further experimentation led him to the use of a latent image, one that only appeared after a development process. Such images allowed much shorter exposure times than had hitherto been possible, which meant that Fox Talbot was no longer forced to use extra large lenses or impracticably small cameras. In 1840, with the freedom to now use more conventionally sized cameras, Fox Talbot introduced his calotype process, which produced paper negatives from which paper positive images could be made.

Although the calotype process was utilised for portraiture, it was used more by landscape photographers. Even after Fox Talbot's announcement, daguerreotype studios continued to flourish and daguerreotypes continued to be the popular medium for portrait photography. That came to an end only when new methods were discovered of producing photographs on glass.

Ambrotypes
If photographic portraiture – and photography in general – were to progress, new methods needed to be found that combined the practicality of the calotype, whose slightly soft negatives could reproduce as many prints as required, with the more sharply defined but one-off and difficult to view image provided by the daguerreotype. That came with the introduction of the wet plate process, announced by English inventor Frederick Scott Archer in 1851.

In place of Daguerre's silver-plated copper-based image and Fox Talbot's paper negative, Archer proposed a glass negative. It was the best solution to date, but it came at a price – its inconvenience of use. The new photographic emulsion Archer had formulated cut exposure times drastically, but the process demanded that the plates were prepared in a darkroom immediately prior to exposure, used in the camera while still wet and then developed immediately after.

Nevertheless, wet plate photography became extremely popular and led to several different types of image favoured by portrait photographers.

FROM TODAY, PAINTING IS DEAD

A sliding box wet plate camera, made by Horne and Thornthwaite.

The first was photographic prints, any number of which could be produced on photographic paper, directly from the glass plate negative.

Equally if not more popular was the ambrotype. This came from a fact that could never have been observed before when images were made on metal plates or paper negatives. It concerned the way that when a glass negative was placed on a dark background, with the side containing the photographic emulsion that had recorded the image uppermost, then the emulsion appeared white, while the black background could be seen through the clear part of the negative. So gluing a piece of black cloth to the back of the plate, or by coating the back with a dark-coloured enamel, turned the negative image into a positive one. The negative plate could no longer be used to make prints, but for the person who wanted a one-off, cheap portrait of themselves, the ambrotype provided the perfect solution.

Ambrotypes, made by the wet place process.

Carte-de-visite and cabinet photographs

Another spin-off of the wet plate process came in the shape of the carte-de-visite. This took the form of a small print, made from the glass plate negative and mounted on a card 2 x 3½ inches, often with an elaborate advertisement on the back for the photographic studio that had provided the service. The prints were most often produced by the albumen process, which used egg whites to bind photo-sensitive chemicals to paper.

The carte-de-visit was another spin-off of the wet plate process.

    The size of carte-de-visites meant that they mostly contained the image of a single person, and only rarely two or more. They were often used as visiting cards and keenly traded among the people who commissioned them.

251

This unusual camera turned carte-de-visites into stamp-like images.

An interesting offshoot of the Victorian craze for different types of photographic portrait came with the advent of cameras made to turn pictures into stamps. One such camera, made by the English Lancaster company in 1896, enabled the user to fit a carte-de-visite into a copying frame, along with a decorative border, and then used nine lenses to photograph it onto a single photographic plate. When the images were contact-printed onto paper with stamp-like perforations, they resembled real postage stamps.

Carte-de-visite portraits later gave way to cabinet photographs, also initially made by the albumen process and mounted on card, but in a different size that measured $4^{1}/_{4}$ x $6^{1}/_{2}$ inches.

Tintypes
In 1856, the wet plate process spawned yet another type of photographic portrait, this time on metal. The process was much the same as that used

for making glass plates, but instead the image was exposed onto a thin sheet of black enamelled iron. The resulting positive image followed the same precept as that of the ambrotype, in this case, the base metal providing the black and the photographic emulsion recording as white. In this way a single direct positive image was formed. The images were known as ferrotypes or tintypes.

The development of faster developing solutions meant that tintypes could be produced very quickly and were much used by itinerant photographers who would photograph people in the street and present them with an image of themselves, still wet, a minute later.

The cheap tintype or ferrotype produced a photographic image on thin metal.

Dry plate photographs

Aside from the cheap and cheerful tintype, photography that used glass plates remained the best bet for sheer quality. Its drawback, of course, was the way it had to be made immediately prior to exposure and developed immediately after. That problem was solved in the 1870s, when English photographer and physician Richard Leach Maddox announced a way to suspend the sensitive photographic emulsion in gelatine and coat it onto a glass plate that could be stored dry, used when needed and

Cabinet photographs spanned the wet plate and dry plate eras.

developed much later. His announcement led to the first factory produced dry plates being introduced in 1878.

Cabinet photographs first seen in the wet plate era now continued to be produced using dry plates, and photography on glass became the standard for use by portrait studios, remaining that way well into the twentieth century. Even with these latest advances, photography in Victorian times was still very much the preserve of the professional or keen amateur. Portraiture – or any other type of photography – was not readily available to the non-expert man in the street. George Eastman was the man who changed all that.

Snapshot photography
Eastman was born in 1854 at Waterville, New York, but at the age of six, his family moved to Rochester. He became interested in photography at the age of twenty-four and soon started to look at ways to make the photographic process simpler, at first by manufacturing his own dry plates, and then by considering what no one had successfully tried before:

coating the photographic emulsion, not on glass, but first onto rolls of paper coated with gelatine. It was this that he used in the first roll film camera, which he launched in 1888 and called the Kodak – the first use of that name. Speaking later about where the name came from, he said:

> The letter K had been a favourite with me – it seems a strong, incisive sort of letter. It became a question of trying out a great number of combinations of letters that made words starting and ending with K. This is not a foreign name or word. It was constructed by me to serve a definite purpose. It has the following merits as a trademark. It is short. It is not capable of mispronunciation. It does not resemble anything in the art and cannot be associated with anything else in the art.

The Kodak was box shaped and had only three controls: a string to tension the shutter, a button to release it and a key to wind the film. It did away with the need for photographers to develop and print their own pictures.

A far more casual form of snapshot portraiture is shown in this picture, taken with the first Kodak.

Instead, the camera came ready-loaded with enough film for 100 exposures. When they had all been shot, the camera and film together were returned to the Eastman works for developing and printing. The camera was reloaded with a new film and returned to the photographer. Five days after that, the owner received the prints, mounted and burnished, together with the negatives.

Suddenly, people who had never before thought of owning a camera began buying and using them for all kinds of snapshot photography. For the first time, photographic portraiture moved out of the studio, into the home and even into the streets, where far more casual and even candid photography of people became both possible and popular.

Photography for the masses had started, and in 1900, one year before the end of Victoria's reign, it was consolidated with a camera whose name was to spell out amateur photography for more than a century. That was the year that Eastman introduced the first Brownie, and the rest, as they say, is history.

CHAPTER 26

Victorian Flying Machines

It has been well documented that the first truly successful attempt at controlled manned flight in a heavier-than-air machine was accomplished by inventors and aviation pioneers Orville and Wilbur Wright. That was in 1903, two years after the death of Queen Victoria. It stands to reason, then, that during her reign, there must have been a lot of budding aviators trying – and failing – to be the first to fly.

Not every Victorian flying machine idea was practical or feasible, as is demonstrated by this speculative illustration from The Graphic *of December 1877. The picture, without further explanation, is simply titled 'A suggestion for a flying machine'.*

Glider pioneers

One of the earliest to make successful repeated flights in a glider was German pioneer Otto Lilienthal, whose machine, built in 1894, was not dissimilar to a modern hang-glider, in which the pilot was suspended from a bar beneath wide wings. Using this, flights were made from both natural hills and one that Lilienthal had built for the purpose near Berlin. Direction of travel was controlled by the pilot shifting his weight from side to side. It was his various successes that earned Lilienthal the name of the Glider King.

Octave Chanute's multi-winged glider.

Prime among other important pioneers was Octave Chanute, a civil engineer born in France in 1832. After working for much of his life in the American railway industry, Chanute turned to studying aviation after he retired in 1883. Being too old to fly by this time, he teamed up with a couple of younger partners named Augustus Herring and William Avery. Together they built and tested a number of ideas for gliders, which led

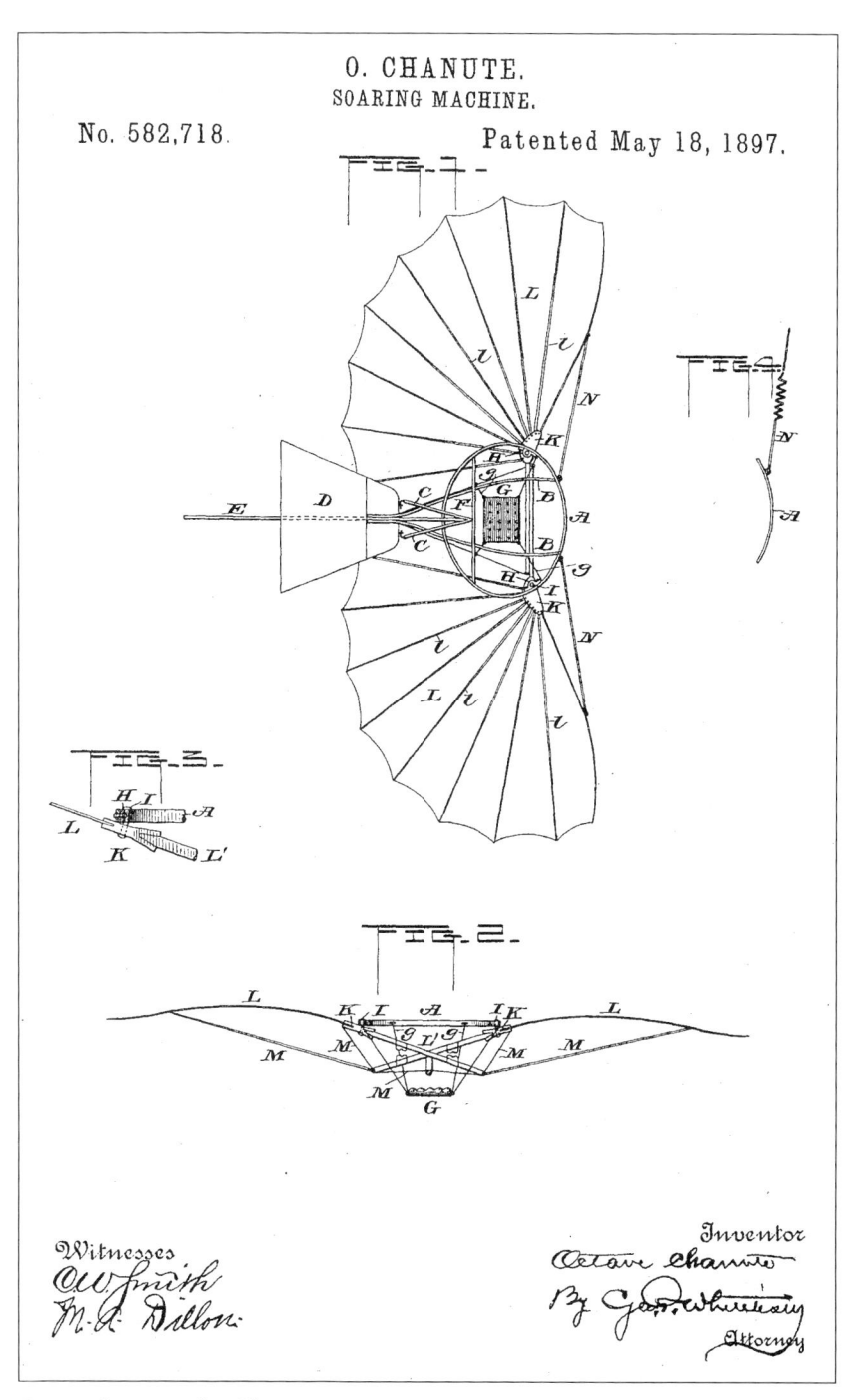

Patent drawings for Chanute's soaring machine.

259

Chanute to at first believe that the way forward was with a glider that used twelve pivoting wings in two tiers of six.

It wasn't a great success, which led Chanute and his partners to taking a more conventional approach with a two-winged glider that was based on an improved version of Lilienthal's original design. In his patent of 1897, Chanute described his device as a soaring machine, stating:

> The object is to imitate the soaring of birds, which is accomplished while the wings are held rigidly still, so far as any flapping movement is concerned. It is necessary, however, to provide for fore-and-aft movement of the wings in order to preserve the equilibrium.

Chanute claimed his improvement on Lilienthal's glider was to provide a stationary seat for the pilot with each wing pivoted in a way that allowed it to move backwards and forwards and so preserve the balance of both the machine and the pilot.

Optimistic aviators like Lilienthal, Chanute and others showed that manned flight was possible, but only if the machine used started from an elevated position from which it glided to earth. Starting from ground level, taking off and travelling through the air was a different challenge entirely, but it was one that was already being taken up by inventors in Britain, France, Germany and America.

Jean Le Bris's Artificial Albatross
Jean Marie Le Bris was a sailor who took his inspiration for a flying machine from watching albatrosses in flight. So he killed one and examined its wings. The glider he built in 1856 was based on the scaled-up shape of an albatross. It had a 50-foot wing span, with hand-operated levers that changed the angles of the wings and foot-operated pedals that changed the position of the tail. He called his craft the *Artificial Albatross*.

The contraption was mounted onto a horse-drawn cart, which was driven along a road and into the wind with Le Bris on board. When he judged that sufficient speed had been reached, he untied the restraining ropes, and according to reports of the day, his machine rose about 300 feet into the air and made a short flight of about 600 feet. One report of the time suggested that as it rose, it took the cart driver with it, tangled from the restraining rope.

Le Bris's machine, based on the way he saw albatrosses fly.

Although the *Albatross* was still essentially a glider, it did have the distinction of starting from ground level and flying upwards, rather than starting at a high level and gliding down. Even so, a second attempt at flying the *Albatross* involved it being launched from a mast about 100 feet above the ground. That flight wasn't so successful. It hit the ground too hard and broke, as did one of Le Bris's legs.

Thomas Moy's Aerial Steamer
In 1874, English engineer Thomas Moy built a machine called the Aerial Steamer. It consisted of a steam engine fuelled by methylated spirits, which drove two huge six-bladed propellers mounted between two wings made from linen stretched across bamboo frames. It was tested, unmanned but tethered, on a circular gravel track around a fountain in the grounds of the Crystal Palace in South London. On the first test run all it did was churn up the gravel. So wooden decking was laid over the gravel and a second attempt made.

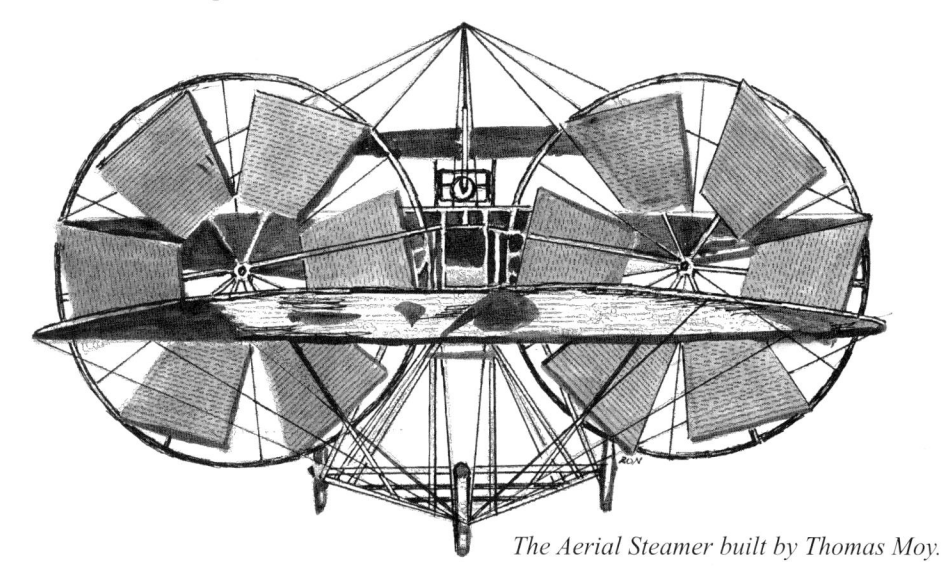

The Aerial Steamer built by Thomas Moy.

THE INGENIOUS VICTORIANS

Moy had calculated that his machine needed to reach 35 miles per hour to lift off, but it never exceeded a speed of 12 miles per hour, and according to Moy, never left the ground. Aviation historian Charles Gibbs-Smith, however, later suggested that it managed to lift off by 6 inches.

Charles Ritchel's gas bag

Getting a man off the ground and into the air for a controlled flight was something addressed by American inventor Charles Ritchel, who was responsible for the curved distorting mirrors popular in fairgrounds of the time and a mechanical toy moneybox, as well as a great many other inventions. His flying machine design involved a brass frame in which the pilot sat, suspended beneath a cylindrical rubber bag of lighter-than-air gas. A propeller at the front of the framework was turned by a crank operated by the pilot. In 1876, Ritchel flew his craft inside an exhibition hall at Philadelphia's Centennial Exposition, and he continued to demonstrate short but moderately successful flights over the next two years.

Charles Ritchel used a bag of lighter-than-air gas to get airborne.

VICTORIAN FLYING MACHINES

The use of a bag of gas to lift the machine and keep it aloft still didn't fully address the challenge of using manpower to lift a flying machine off the ground and keep it in the air while in motion.

Alexandre Goupil's sesquiplane

French engineer Alexandre Goupil was one person who took up that challenge in 1883. His machine took the form of a large rounded body that housed a steam engine to drive a propeller at the front, with a rudder hanging from the back. A wing protruded from each side of the housing, with a total wingspan across the two of nearly 20 feet. Below the housing, a structure allowed a man to stand between two landing wheels, with his feet resting on pedals attached to the front wheel, with a slightly smaller one at the rear. His machine was called a sesquiplane.

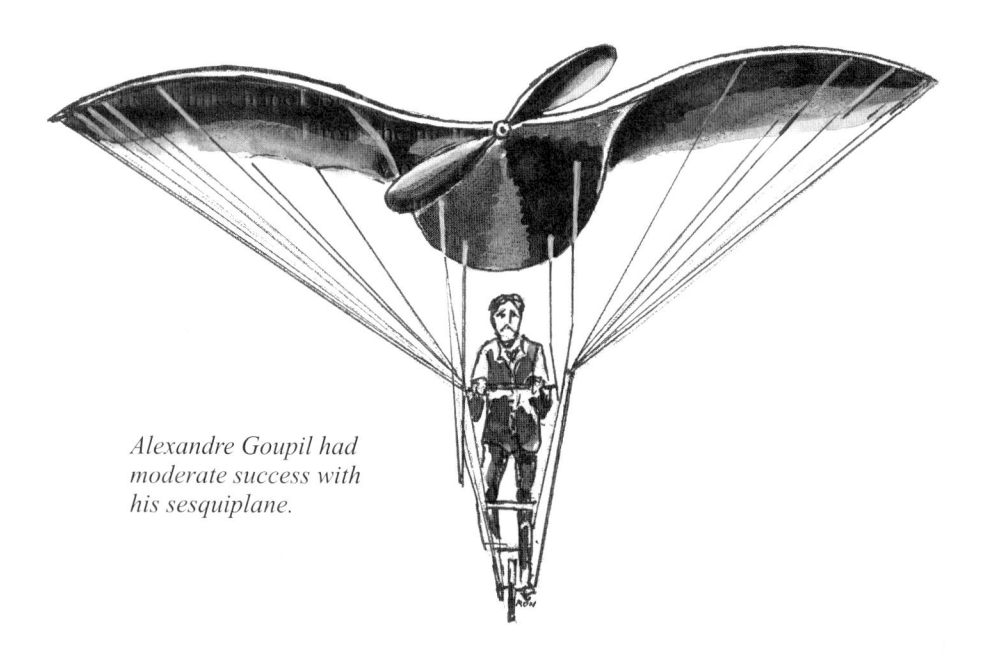

Alexandre Goupil had moderate success with his sesquiplane.

There's no evidence to suggest that the machine was ever tested with its steam engine, but an unpowered version was known to have lifted into the air with two men on board, in a wind of about 14 miles per hour.

Clement Ader's bat planes

If claims by French inventor Clement Ader were to be believed, he had rather more success with steam-powered flying machines. He made two, called the Ader Éole and the Ader Avion No. 3, its two predecessors having been damaged in trials and experiments. Both machines had wide wingspans with steam-driven propellers, and bore a certain resemblance to gigantic bats.

Later in life, many of Ader's somewhat exaggerated claims were disproven, but it is pretty much accepted that, in 1890, he did manage to make a very short manned flight in the Ader Éole, which eventually earned him the title, among his fellow countrymen, of the Father of Aviation.

Hiram Maxim's gigantic biplane

In that same year, American-born but nationalised English inventor Hiram Maxim built a wind tunnel for his experiments, and in 1891 he patented an enormous biplane. Maxim's machine was 145 feet long, with a wingspan of 105 feet and two steam-driven propellers. Later, he added more wings. Because it was not easy to control and would be aerodynamically unstable, Maxim built his machine more for research purposes than for true flight. For that reason, it was made to run along railway-like tracks and restrained to prevent lift-off. Nevertheless, in 1894, it broke one of its retaining rails and did indeed lift off, becoming

Hiram Maxim's giant flying machine.

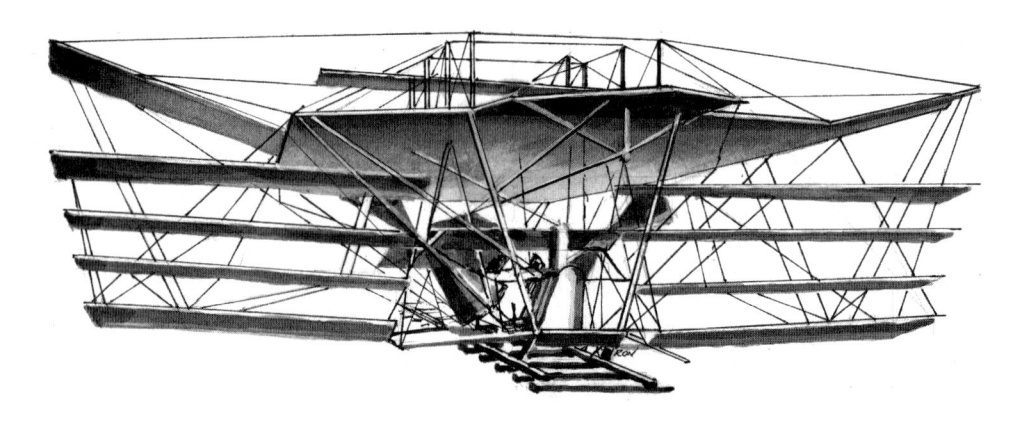

damaged in the process. Maxim then abandoned his experiments until later in the twentieth century, when he switched from steam to internal combustion engines for his experiments in manned flight.

Despite the efforts of these and other Victorian inventors, true manned flight in a heavier-than-air machine didn't become a reality until the Wright brothers' famous first flight in 1903, the first lasting only twelve seconds and covering no more than 120 feet. Three other flights were made on the same day, the best covering 842 feet in fifty-nine seconds. It was two years into the Edwardian age, but might never have happened had it not been for the experimentation and perseverance of so many brave innovators who risked life and limb attempting to fly during the Victorian era.

CHAPTER 27

The People Who Invented Christmas

Before the Victorian age, Christmas as it is known today did not exist. The way it is now celebrated is down to many factors that came together in the Victorian era. They included a book by the author Charles Dickens, a new-found wealth among many people, Pagan festivals, Dutch migrants to America, the emergence of a reliable postal service, and Prince Albert, Queen Victoria's consort.

Although Christians obviously acknowledged their belief in Jesus as the son of God, it never occurred to them to celebrate his birth. For about a century after the beginnings of Christianity, debate raged in the church and beyond about who exactly Jesus was, what he represented and what form he actually took. Eventually it was decided that it might help if agreement might be reached on his date of birth. Various dates were considered and juggled by theologians who eventually announced that Jesus had been conceived during the spring equinox and born on 25 December.

Even so, until the nineteenth century, 25 December was rarely celebrated in any great way in Britain, and businesses certainly didn't see it as a reason for giving employees time off work. What celebrations did take place were linked more to old Pagan festivals: decorating houses with holly and ivy, and a time for feasting. Even so, most saw New Year's Day as more appropriate for celebration and the exchange of gifts.

Christmas trees
All that changed when Victoria came to the throne in 1837 and married her first cousin, Prince Albert of Saxe-Coburg and Gotha, in 1840. It was his influence, more than most, that influenced the start of the Christmas tradition as it is known today, and once it began, it all took off rather quickly.

Victoria, Albert and five of what would eventually become nine children introduce the idea of the Christmas tree in an 1848 issue of The Illustrated London News. *The caption stated: 'The pretty German toy, a Christmas tree.'*

Prince Albert's childhood in Germany involved Christmas celebrations around a decorated tree, a tradition that he introduced to Britain. An engraving that appeared in *The Illustrated London News* in 1848 showed Victoria, Albert and children gathered around a tree, on which there were decorations and candles, and beneath which there were toys. It was the first time that the concept of a Christmas tree hit the public in any significant way, and very soon, many rather less royal families, keen to emulate the Royals, were introducing decorated trees and gifts into their own homes to celebrate the growing interest in a Christmas festival.

Henry Cole's first Christmas card.

Christmas cards

At the same time, the craze for Christmas cards began to emerge. The origins for that had begun a few years before with Sir Henry Cole, an English civil servant and innovator in the fields of commerce and education. In 1843, he commissioned an artist to produce the first commercial Christmas card. It depicted a fairly affluent looking family celebrating and drinking together, surrounded by images of what appeared to be poorer families, a great deal of intertwining vines and a message to say 'A Merry Christmas and A Happy New Year to you'. There were spaces at the top and bottom for the user to write in 'To' and 'From' names.

The card sold for one shilling (5p), which was rather beyond the means of the general public, but the idea caught on, particularly among children, who were subsequently encouraged to make their own cards. Two more factors then contributed to the growing popularity of Christmas cards: new advances in printing technology and the introduction of the halfpenny postage rate. By 1880, the production and selling of Christmas cards became a major industry and it was estimated that more than 11 million cards were produced in that year alone.

THE PEOPLE WHO INVENTED CHRISTMAS

Christmas crackers

Another step in the popularisation and development of the traditional Christmas came in 1848, with the introduction of the first Christmas crackers. They were the brainchild of confectioner Tom Smith. In 1840, he went to Paris, where he saw, for the first time, the French bon-bon – a sugared almond wrapped in a small piece of tissue paper. He brought the idea home with him and introduced his version of the bon-bon to Victorian London, soon discovering they sold best at Christmas.

Bringing home the yule log – a Victorian Christmas card.

Another Victorian card, this time depicting skating frogs.

THE INGENIOUS VICTORIANS

Developing the idea further, he added a small love motto to each of his bon-bons. Then, to encourage further sales around Christmas, he added a simple mechanism inspired by the crackle of a log on a fire. After some experimentation, he devised a mechanism that set off a very small and safe explosion as the two halves of the now somewhat larger paper wrapping were pulled apart. Thus was born the Christmas cracker.

Christmas decorations and presents

Meanwhile, the tradition of decorating a house at Christmas began to take on new dimensions, as the traditional Pagan-inspired methods of adding sprigs of greenery to walls blossomed into far more adventurous styles of decorations, many inspired by women's magazine articles of the day.

Encouraged by Victorian engravings showing Victoria and Albert with Christmas gifts for their children, present giving among the public now began to shift from New Year, a week back to Christmas Day.

Once the craze for Christmas presents had begun, it didn't take long for major retailers to cash in, as this 1888 advertisement for Mappin & Webb in London shows.

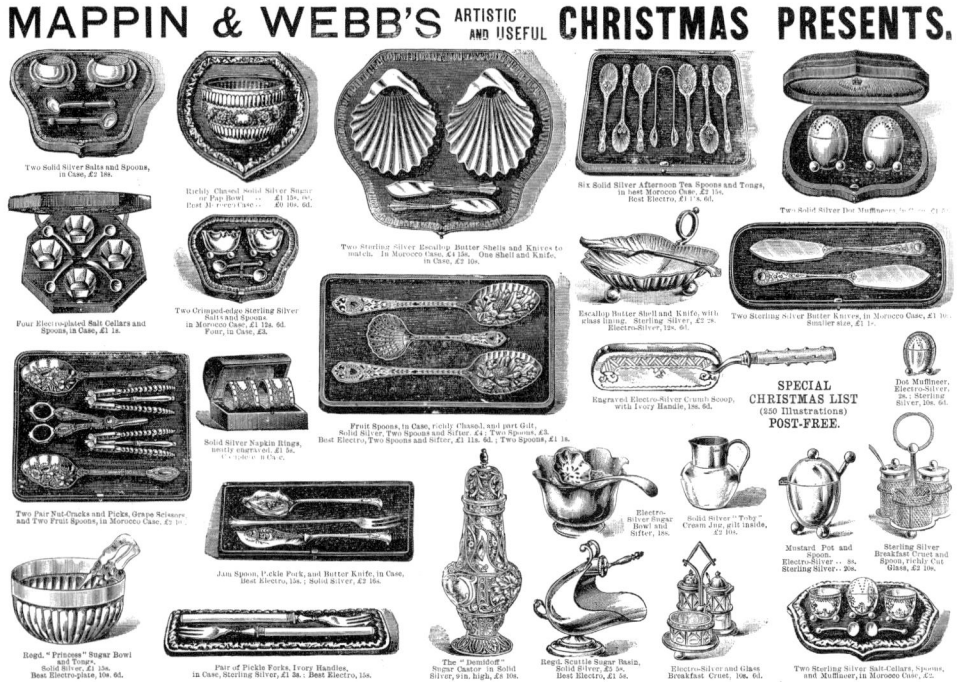

A Victorian version of Father Christmas by illustrator Jenny Nyström (1854–1946).

Father Christmas

Santa Claus and Father Christmas were originally two different people, both of whom were around before the Victorian age.

Father Christmas was part of the midwinter festivals held in Britain, a pagan figure who represented the coming of spring. He was dressed in a green, rather than red, cloak with a hood, wearing a wreath of holly, and was more associated with feasting and drinking than the giving of presents.

Meanwhile in America, early settlers rejected the idea of an English Father Christmas, but when Dutch immigrants arrived, they brought with them the legend of Sinterklaas, a dialectal mispronunciation of Saint Nicholas, the historical Greek bishop known for bestowing gifts. In a poem thought to be by Clement Clarke Moore, an American professor of Greek Literature, Saint Nicholas was later described as a man dressed in fur, arriving in a sleigh pulled by reindeer.

THE INGENIOUS VICTORIANS

By the time all these various legends arrived in Victorian England, Sinterklaas had become Santa Claus, soon to be confused with Farther Christmas, the two coalescing into one, now wearing a red cloak, driving a flying sleigh and distributing gifts to children.

Christmas dinners and carols
Feasting at this time of year was nothing new, but now it became more regularised as affluent people added turkeys in place of beef and goose, which had previously been more popular. As was so often the case, what the rich did one day, the poorer classes did the next, and soon turkey became the traditional Christmas food.

Carols were not new either, but they were revived in the Victorian era, with new words to old tunes, and the first official book of carols was soon published.

Now, as businesses started to benefit from a growing affluence in society, bosses began to reluctantly give their employees time off, and Christmas became a day of eating, celebrating and the bringing together of families.

The concept of Christmas being a time for family gatherings was greatly perpetuated by Charles Dickens, who in 1843 wrote *A Christmas Carol*. The author wrote the book in six weeks, starting in October of that year and finishing it in November, in time for publication on 19 December. The tale of the skinflint miser Ebenezer Scrooge being shown the error of his ways by the visitation, on Christmas Eve, of three ghosts, was an immediate and huge success, selling more than 6,000 copies in a week. Within six weeks, it had been adapted as a London stage play, which ran for more than forty nights before transferring to a theatre in New York.

Charles Dickens didn't actually invent Christmas as we know it today, but his hastily written book *A Christmas Carol* went a long way towards establishing the template for the Victorian family Christmas, and one that still remains today.

A typical family Christmas, as visualised in an issue of The Illustrated London News *of 1894.*

CHAPTER 28

An Amazing Memorial and the End of an Era

Prince Albert, husband and consort to Queen Victoria, died of typhoid in December 1861 at the age of forty-two. Despite his achievements – not least his involvement with The Great Exhibition – he was not over popular with the Victorian public, and during his all too brief lifetime he made it clear that he was not in favour of monuments or statues being erected in his honour. He was, however, the love of Queen Victoria's life, and when he died, she thought otherwise. As a result, London acquired perhaps its most ornate, extravagant, grandiose and totally over-the-top memorial.

At Queen Victoria's request, the memorial was not meant as a tribute to Albert's achievements as such, but more a general memorial to his memory. In 1862, a year after Albert's death, Lord Mayor of London William Cubitt appointed a committee to raise funds, and for a design to be prepared for the Queen's approval. The winning design came from English architect Sir George Gilbert Scott. Although he began his career designing workhouses, Scott was best known for his association with churches and cathedrals, and for the design of St Pancras Station.

Scott's design reflected his interest in the Gothic Revival style. He was quoted as saying: 'There can, indeed, be no doubt that the public expect a monument of great and conspicuous magnificence.' Which is exactly what the public got.

Work began on what was officially known as the Prince Consort National Memorial in 1864. It was a complex structure supported by a massive steel cross, with an undercroft of brick arched tunnels forming vaults in the foundations, similar to the crypt of a church. Today, these tunnels can still be accessed via a manhole in front of the memorial.

The builder was Sir John Kelk, who was responsible for building the Victoria Station and Pimlico Railway. The memorial's cost of £120,000 was

AN AMAZING MEMORIAL AND THE END OF AN ERA

The Albert Memorial as it was in Victorian times.

met by public subscription, and Kelk guaranteed to carry out the work at cost price without any personal profit to himself, whether directly or indirectly.

Over the coming years, the memorial grew as huge pieces of stone were assembled on site, with an overhead crane moving the sections into place. As it grew taller, scaffolding was erected for work to be carried out on the highest parts of the structure.

The original design for the centrepiece statue of Prince Albert came from Italian-born French sculptor Carlo Marochetti, but his ideas were deemed unsatisfactory. When he died, the work was taken over, in 1868, by Irish sculptor John Henry Foley, who had already been involved with some of the groups of statues that surrounded the memorial.

Queen Victoria officially unveiled the Albert Memorial in 1872, but at that stage it still lacked one important aspect – the statue of Prince Albert. That wasn't put in place until 1875, three years after the official opening and twelve years after work had begun.

Over the following century, the Albert Memorial gradually deteriorated. But in the 1990s it was slowly and methodically restored for

The fully restored Albert Memorial as it is today, very much the way it would have been seen by the Victorians.

AN AMAZING MEMORIAL AND THE END OF AN ERA

a new unveiling in 1998. The memorial standing in Kensington Gardens today, complete with a shining central statue, is very much the way it would have been seen by the Victorians.

The Albert Memorial is 176 feet high. At the four corners of its base stand sculptures that represent Europe, Asia, Africa and the Americas. From here, flights of marble steps rise on each side to the base of the monument, where four more statues, one at each corner, represent Victorian arts and sciences: agriculture, commerce, engineering and manufactures.

Standing at the base, clockwise from top left: statues recognising Europe, Asia, Africa and the Americas.

Clockwise from top left: agriculture, commerce, engineering and manufactures are represented by further statues.

Part of the elaborate frieze that surrounds the monument.

AN AMAZING MEMORIAL AND THE END OF AN ERA

An elaborate frieze that shows 169 famous figures surrounds the central part of the memorial. Eminent architects are shown on the north side, musicians and poets are depicted on the south, painters are on the east, and sculptors are on the west.

Above this sits the gilt bronze statue of Prince Albert, robed as a Knight of the Garter, with the catalogue for The Great Exhibition on his knee as he looks south towards the Royal Albert Hall.

Above the statue is a canopy featuring mosaics, representing poetry, painting, architecture and sculpture. Around this canopy, split into four sections, runs the message: 'Queen Victoria And Her People – To The Memory Of Albert Prince Consort – As A Tribute Of Their Gratitude – For A Life Devoted to the Public Good.'

The statue of Albert shows him robed as a Knight of the Garter and holding a catalogue for The Great Exhibition on his knee.

Within niches in the canopy, statues stand representing astronomy, geology, chemistry, geometry, rhetoric, medicine, philosophy and physiology.

More statues stand near the top of the canopy's tower to depict moral and Christian values: faith, hope, charity, humility, fortitude, prudence, justice and temperance. Above these, gilded angels raise their arms to the sky and a gold cross stands at the pinnacle of this most amazing memorial.

Queen Victoria lived for a further forty years after Prince Albert's death, and wore black every day in his memory. She died in 1901 on the Isle of Wight at her summer home, Osborne House, a place which itself was a tribute to Victorian ingenuity. Rather than being an extension of an existing house, which was the case with so many nineteenth-century country houses, it was intended and purpose-built for specific family use, unlike the way other royal residences of the time were designed.

Osborne House on the Isle of Wight, where Queen Victoria died in 1901.

    The entire estate today covers nearly 350 acres, which still include a small Swiss Cottage with its own gardens, once tended by Victoria's nine children. Nearby is the Queen's own private beach, where she regularly bathed and her children leaned to swim.

    Osborne House was designed by Prince Albert, in the Italian Renaissance style. He said that the views from the house, across the Solent, reminded him of the Bay of Naples, and before her death, Queen Victoria said of this area: 'It is impossible to imagine a prettier spot.'

    Her death in this remarkable house marked the end of one of the most innovative and ingenious eras in British history.

Bibliography

Books
A Brief Biography of Thomas Alva Edison, Thomas Alva Edison Foundation.
A Centenary History of The Kodak, The Kodak Museum.
Descriptive Illustrated Catalogue of the Sixty-Eight Competitive Designs for The Great Tower For London, originally published in 1890, reprinted by Isha Books.
Gernsheim, Helmut and Alison, *The History of Photography*, Thames & Hudson.
Gibbs-Smith, C.H., *The Great Exhibition of 1851, a Commemorative Album*, HM Stationery Office.
Hadfield, Charles, *Atmospheric Railways*, David & Charles.
Howarth, Patrick, *The Year is 1851*, White Lion.
Kentley, Eric (ed), *The Brunels' Tunnel,* The Brunel Museum.
Mansfield, Ian, *London's Lost Pneumatic Railways*, Kindle Edition.
Rochford, Caroline, *Great Victorian Inventions*, Amberley.

Journals
Popular Science Monthly, August 1875.
Railroad, 1901.
The Engineer, various issues from 1837–1901.
The Graphic, various issues from 1837–1901.
The Illustrated London News, various issues from 1837–1901.
The Scientific American, various issues from 1837–1901.

Articles and other sources
Correspondence from J.A. Anderson to photographic publishers in 1900.
'Photography Genius: George R. Lawrence & "The hitherto impossible"', Janice Petterchak in the *Journal of the Illinois State Historical Society.*
The Largest Photograph in the World of the Handsomest Train in the World. Pamphlet distributed by P.S. Eustis, General Passenger Agent, Chicago in 1901.
'The Pneumatic Despatch Company's Railways'*:* Notes from a lecture by Charles E. Lee, FCIT, given at the Science Museum in 1973.

Picture Credits

Page 8: © John Wade.
Page 9: General Post Office of the United Kingdom of Great Britain and Ireland. (Picture from Wikimedia Commons.)
Page 10: © John Wade.
Page 12: Upper, © John Wade; lower, Catskill Archive, originally published in *Scientific American*, 1886. (Picture from Wikimedia Commons.)
Page 13: From *The Illustrated London News*, 1884.
Page 15: From *The Illustrated London News*, 1851.
Page 17: © Victoria and Albert Museum, London.
Pages: 18–34: From *The Illustrated London News*, 1850/1851.
Page 36: © John Wade.
Page 37–8: From *The Illustrated London News*, 1854.
Pages 39–40: From contemporary post cards.
Page 41: Upper and lower, © John Wade.
Page 42: From a contemporary postcard.
Pages 43–5: © John Wade.
Page 47: © Victoria and Albert Museum, London.
Page 51: Courtesy of Bernard Volk.
Page 52: © John Wade, with thanks to Brighton Toy and Model Museum.
Page 53: Original painting by Richard Shaw.
Page 54: Courtesy of Grange Museum, Rottingdean.
Pages 55–6: From contemporary postcards.
Page 58: From the Douglas d'Enno collection.
Page 59: © John Wade.
Page 60: Sketch by John Jabez Edwin Mayall. (Picture from Wikimedia Commons.)
Pages 62–4: From *The Illustrated London News,* 1851.
Page 67: Picture from Wikimedia Commons.
Pages 69–73: From *The Largest Photograph in the World of the Handsomest Train in the World* – pamphlet distributed in 1901 by P.S. Eustis, General Passenger Agent, 209 Adams Street Chicago. Courtesy of Charlie Kamerman.
Page 75: From *The Illustrated London News*, 1864.
Page 76: Upper, © Nick Catford, Subterranea Britannica; lower left and right, from *The Illustrated London News*, 1882.
Page 77: Upper, from *The Illustrated London News*, 1882; lower, © Nick Catford, Subterranea Britannica.

PICTURE CREDITS

Page 78: Upper, courtesy of Nick Catford, Subterranea Britannica; lower, © Nick Catford, Subterranea Britannica.
Pages 79–81 © Nick Catford, Subterranea Britannica.
Page 84: From a catalogue first published 1890.
Page 85: From *The Engineer*, 1890.
Page 87: From *The Graphic*, 1894.
Page 88: © TfL, from the London Transport Museum collection.
Pages 89–90: From *The Graphic*, 1894.
Pages 91–3: From contemporary postcards.
Page 94: From a Victorian programme for Blackpool Tower.
Page 95: Courtesy of Blackpool Entertainment Company Ltd.
Page 98: Upper and lower, from *London*, published, 1842.
Page 99: From *Old and New London*, published 1880.
Page 100: From a contemporary book illustration.
Pages 101–102: © John Wade.
Pages 104–105: From *The Engineer*, 1893.
Page 106: Upper and lower, From *Black and White*, 1891.
Page 107: Upper, from *Black and White*, 1891; lower, from a contemporary postcard.
Pages 108–10: From *The Graphic*, 1894.
Page 111: © Tower Bridge, City of London Corporation.
Page 112: Upper and lower, from contemporary postcards.
Page 114: Upper and lower, from contemporary postcards.
Page 115: From *The Illustrated London News*, 1889.
Page 116: From a contemporary postcard.
Pages 119–20: Courtesy of John Jenkins, www.sparkmuseum.com.
Page 121: From a contemporary advertisement.
Pages 122–3: Courtesy of Caroline Rance, www.thequackdoctor.com.
Page 124: From a contemporary advertisement.
Page 128: Courtesy of Auction Team Breker, Germany.
Page 129: Courtesy of Professor John Hannavy.
Page 130: Upper, from a contemporary advertisement; lower, © John Wade.
Page 131: Upper, courtesy of Christie's Images Ltd; lower, © Nigel Richards.
Page 132: From a contemporary advertisement.
Page 133: Upper, John Wade collection; lower, © Lew Steinfeld.
Page 134: John Wade collection.
Page 135: From Wikimedia Commons.
Pages 136–7: From *Popular Science Monthly*, 1875.
Page 139: From Wikimedia Commons.
Page 141: From *The Subterranean Railway*, 1855.
Page 143-5: From *The Illustrated London News*, 1860/1861/1862.

Page 146: From *Old and New London*, 1880.
Page 147: Upper, from *The Illustrated London News*, 1860; lower, © TfL from the London Transport Museum collection.
Page 148, Left and right, © John Wade.
Page 149: © TfL from the London Transport Museum collection.
Page 152: © Geoff Howard.
Pages 154–8: Courtesy of the United States Patent and Trademark Office.
Page 159: From a contemporary advertisement.
Pages 162–9: From *The Illustrated London News*, 1861/1845.
Page 170: © John Wade.
Page 173: From *The Illustrated London News*, 1877.
Pages 175–6: From *The Graphic,* 1877/1878.
Page 177: From a contemporary postcard.
Pages 178–9: © John Wade.
Page 180: Courtesy of the Charles Edison Fund.
Page 182: Courtesy of the United States Patent and Trademark Office.
Pages 183–4: From a contemporary book.
Pages 185–6: © John Wade.
Page 187: Courtesy of United States Library of Congress's Prints and Photographs division, digital ID cph.3c24124.
Page 188: © Andreas Praefcke. (Picture from Wikimedia Commons.)
Page 189: © John Wade.
Page 190: From Wikimedia Commons.
Page 192: Courtesy of George Eastman House.
Page 194: Courtesy of Alan Acres.
Page 195: From a contemporary advertisement.
Page 196: Upper, from Wikimedia Commons; lower, courtesy of Christie's Images Ltd.
Page 198, Left and right, from John Wade collection.
Page 201: From *Brockhaus' Conversations-Lexikon*, 13th edition, 1887.
Page 202: Courtesy of the United States Patent and Trademark Office.
Page 203: From an original poster by Karl Edwards.
Page 204: From *The Boys Own Paper*, 1897.
Page 205: Courtesy of the United States Patent and Trademark Office.
Page 206: From *The Engineer*, 1869.
Page 207: Upper, from *The Tricyclists Indispensible Annual and Handbook*, 1883; lower, from a contemporary advertisement.
Pages 208–209: From Wikimedia Commons.
Page 210–11: Courtesy of the United States Patent and Trademark Office.
Page 213: Donaldson Lithographic Company. (Picture from Wikimedia Commons.)

PICTURE CREDITS

Page 215: From *The Pneumatic Despatch*, 1868.
Page 216: From a news press wire service photograph of 1865.
Page 217: From *The Illustrated London News*, 1864.
Page 220: From *The Engineer*, June 1877.
Page 221: From *Scientific American*, 1870.
Page 223: Courtesy of the United States Patent and Trademark Office.
Page 225: From a contemporary French postcard by Marius Bar.
Page 226: World Imaging. (Picture from Wikimedia Commons.)
Page 229: From a contemporary postcard.
Page 230: From *The Graphic*, 1887.
Page 231–3: From *The Illustrated London News*, 1858.
Page 234: From *The Graphic*, 1887.
Page 235: © John Wade.
Page 237–40: From *The Illustrated London News*, 1851.
Page 241–2: © John Wade.
Page 243: © Philip Merryman.
Page 245: Artwork by Ron Holloway
Page 246: Upper and lower, from John Wade collection.
Page 247: From the collection of Jean-Christophe Badot.
Page 249: © John Wade.
Page 250: Courtesy of Bob White.
Page 251–4: © John Wade.
Page 255: From John Wade collection.
Page 257: From *The Graphic*, 1877.
Page 258: Artwork by Ron Holloway.
Page 259: Courtesy of the United States Patent and Trademark Office.
Pages 261–4: Artwork by Ron Holloway.
Page 267: From *The Illustrated London News*, 1848.
Page 268: John Callcott Horsley. (Picture from Wikimedia Commons.)
Page 269: Upper and lower, Nova Scotia Archives. (Pictures from Wikimedia Commons.)
Page 270: From *The Illustrated London News*, 1888.
Page 271: Minnesota Historical Society. (Picture from Wikimedia Commons.)
Page 273: From *The Illustrated London News*, 1894.
Page 275: From a contemporary postcard.
Pages 276–80: © John Wade.

Index

Acres, Birt, 189, 193–4
Ader, Clement, 264
Aeolia harp, 138, 140
Aerial Steamer, 261
Albert Memorial, 274–9
Albert, Prince, 15, 23, 24, 26, 36, 49, 266–7, 274, 280
Alexandria, Queen, 134
Alton, F.G., 11
Ambrotypes, 248–9
Analytical engine, 9
Archer, Frederick Scott, 248
Armat, Thomas, 193
Artificial Albatross, 260
Atmospheric railways, 48–9, 141
Babbage, Charles, 9
Baker, Benjamin, 90, 174, 177
Barnett, Frederio, 103
Barry, Charles, 228
Barry, John Wolfe, 104,
Bat plane, 264
Baudre, Honoré, 140
Bazalgette, Joseph, 103, 163, 170, 171
Beach, Alfred, 221
Beaumont, Frederick, 76
Bell, Alexander Graham, 7, 9, 185
Benn, John, 117
Bentham, Samuel, 103
Berliner, Emil, 187
Bicycles, 203–13
Bicyclette, 204
Big Ben, 228–35
Biograph, 197
Birtac, 194
Blackpool Tower, 92–5
Blackpool Tower Company, 92
Bloch's Photo Cravate, 131
Bobeth & Schulenberg, 159,160
Book cameras, 131
Brighton and Rottingdean Seashore Electric Railway, 51–9
British Empire Exhibition, 91
Brontë, Charlotte, 35
Brownie, 256
Brunel, Isambard Kingdom, 96, 99
Brunel, Marc, 96–101

Cabinet photographs, 250, 254
Calotypes, 246–8
Carstensen, Georg, 44
Carte-de-visite, 250–2
Caunt, Benjamin, 233
Channel Tunnel, 10, 74–82
Chanute, Octave, 258–60
Christmas
  cards, 268–9
  carols, 272
  crackers, 269
  decorations, 270
  dinner, 272
  presents, 270
  trees, 266–7
Cinématographe, 195–7
Cleopatra's Needle, 172–9
Cocker, Doctor, 92–3
Cole, Henry, 24, 268
Computers, 9
Crooks, Will, 117
Cros, Charles, 187
Crystal Palace, 17–22, 37–45, 61, 217, 236, 261
Crystal Way, 49
Daddy Long-Legs, 51, 53, 55
Daguerre, Louis, Jacques, Mandé, 244
Daguerreotypes, 244–6
Dandy Charger, 200
Delaroche, Paul, 245
Denison, Edmund Beckett, 229–31
Dent, Edward, 230
Dickens, Charles, 127, 266, 272
Difference engine, 9
Dixon, John, 174
Donisthorpe, Wordsworth, 189, 191
Dribble, Cornelis, 222
Dry plate photographs, 253
Duell, Charles, 11
Duncan, Francis, 115
Durham, Joseph, 36
Eagle bicycle, 203
Eastman, George, 191, 192, 254
Edison Electric Light Company, 180
Edison, Thomas Alva, 11, 83, 180–8, 189, 192
Eiffel, Gustave, 83, 85
Eiffel Tower, 10, 83, 88, 92

INDEX

Electro-therapy, 118–21
Electropathic & Zander Institute, 123
Elgar, Edward, 91
English, Thomas, 76–7
Father Christmas, 271
Fearnaught, Albert, 154
Ferrotypes, 252–3
Fibre optics, 9
Field, George, 153
Flying Yankee Velocipede, 206
Foley, John Henry, 275
Friese-Greene, William, 189–91
Gladstone, William, 146
Gloucester Railway Carriage and Wagon Company, 53
Gordon, Charles, 115
Goubet, Claude, 222
Goubet I, 222–5
Goubet II, 225–6
Goupil, Alexandre, 263
Gramophone, 187–8
Grant, John, 171
Graphophone, 185–6
Gray, Elisha, 7
Great Conservatory, 16
Great Exhibition, 21, 22, 23–36, 42, 61, 236, 238, 274
Great Stink, 161, 163
Great Tower For London, 85
Great Victorian Way, 46–50
Greenwich Mean Time, 240, 242
Gutsmuth, Adolf, 153
Gye, Frederick, 49
Gymnote, 226–7
Hall, Benjamin, 233
Handbag cameras, 133
Harness, Cornelius Bennett, 121, 122
Harness's Electric Corset, 122–3
Henman, William, 85, 87
Hill, Roland, 9, 24
Holmes, Sherlock, 127
Hotchkiss, Arthur, 210
Hutton, Charles, 116
Hysteria, 125–6
Illustrated London News, The, 30
International Tower Construction Company, 90
Jazz Singer, The, 199
Jeffery, Thomas, 205
Jenkinson, Robert Banks, 174

Jones, Horace, 103
Kabrich, Chas, 212
Karnicki, Michael, 153
Kastner, Frédéric, 135–40
Kelk, John, 274
Kelvin, Lord, 11
Kinesigraph, 191
Kinetic Camera, 194
Kinetograph, 192–3
Kinetoscope, 194
Kinora, 198–9
Klein, Herman, 184
Knight, John Peake, 11
Kodak, 255–6
Koh-i-Noor diamond, 34
Krebbs, Arthur, 226
Krichbaum, John, 156, 157
L'Exposition Universelle, 83
Lallement, Pierre, 200, 202
Lardner, Dionysius, 11
Lawrence, George, 67–8
Lawson, Henry, 204
Le Bris, Jean Marie, 260–1
Le Prince, Louis, 189, 191
Leinster Gardens, 148
Lilienthal, Otto, 258
Lithophone, 140
London Association for the Prevention of Premature Burial, 152
London Pneumatic Despatch Company, 216
London Underground, 50, 141–50
Lumière, Auguste and Louis, 195
Macmillan, Kirkpatrick, 200
Maddox, Richard Leach, 253
Mammoth, 67–73
Mappin & Webb, 270
Marconi, Guglielmo. 7
Marey's Gun Camera, 129
Mathieu-Favier, Albert, 74
Maxim, Hiram, 264
Maxwell, James, 93
Medical Battery Company, 121
Meigs, Joseph, 12
Metropolitan Railway, 142
Metropolitan Tower, 85
Midland Railway, 16
Monster Globe, 60–6
Morecambe Tower, 92
Mosely, William, 49

287

Mottray, Henry, 74
Mousetrap cameras, 248
Moy, Thomas, 261–2
Murders in the Rue Morgue, 127
Mutoscope, 197-198
New Brighton Tower, 91
Osborne House, 279–80
Paleophone, 181
Paxton, Joseph, 15–16, 22, 37, 44, 45, 46, 87
Pearson, Charles, 141–2, 146
Penny Black, 9
Penny Farthing, 203
Phonautograph, 181
Phonograph, 181–5, 191, 192
Photophone, 9
Pillar boxes, 9
Pinkerton's Detective Agency, 127
Pioneer, 52, 53, 57, 59
Pneumatic railways, 12, 214–21
Poe, Edgar Alan, 127, 151
Postage stamps, 9
Preece, William, 11
Premature Burial, 151
Pugin, Augustus, 231
Punch, 22, 60, 61
Pym, John, 49
Pyrophone, 135–40
Radio, 7
Railroad Velocipede, 210
Rammell, Thomas, 214, 217, 219–20
Redl, Carl, 156
Ritchel, Charles, 262
Road and River Cycle, 212
Rosebury, Lord, 116
Royal Observatory, 240
Safety bicycles, 205–206
Safety coffins, 151–60
Santa Claus, 271
Scott, George Gilbert, 124, 274
Sesquiplane, 263
Sewage system, 50, 161–71
Shepherd, Charles, 238–43
Singing flames, 136
Smee, Alfred, 239
Smith, Robert, 85, 87
Smith, Timothy Clark, 152–3
Smith, Tom, 269–70
Snapshot photography, 254
Snow, John, 162

Society for the Prevention of People Being Buried Alive, 151
South Kensington Pneumatic Railway, 219–20
Squires, William James, 116
St. Paul's Cathedral, 21, 39
Statham, Henry Heathcote, 113
Stewart, MacLaren and Dunn, 88
Stirn's Patent Concealed Vest Camera, 130
Talbot, William Henry Fox, 246–8
Taphophobia, 151, 152
Tarberger, Johann Gottfried, 155
Telephones, 7, 11
Thames Tunnel, 96–103
Theme parks, 10
Thirkell, Angela, 57
Thomé, Aimé, 74
Thompson's Revolver Camera, 128
Thutmose III, 172, 179
Tintypes, 252–3
Toolen, Johan, 153
Topsy-Turvy Railway, 39
Tower Bridge, 103–13
Traffic lights, 11
Transatlantic Times, 8
Tuke, Charles, 93
Velocipedes, 200–203
Victor Rotary Tricycle, 207
Victoria, Queen, 7, 9, 10, 15, 22, 24, 25, 42, 74, 97, 127, 140, 150, 174, 179, 199, 232, 257, 279, 280
Vitascope, 193
Volk, Magnus, 51
Vulliamy, George John, 179
Walking stick cameras, 133–4
Watch cameras, 132, 133, 134
Watkin, Edward, 74, 75, 80–1, 83, 90. 150
Watkin's Folly, 85
Watkin's Stump, 85
Watkin's Tower, 85
Wembley Park Tower, 85
Wembley Stadium, 91
Wembley Tower, 85
Wessheimer, Wendelin, 138
Western Union, 11
Wilson, William James Erasmus, 11, 174, 176
Wireless telegraphy, 8
Woolwich Ferry, 113–17
Wright, Orville and Wilbur, 257
Wyld, James, 60–6